BASIC GAMBLING MATHEMATICS

MATHEMATICS

*The Numbers
Behind the Neon*

BASIC GAMBLING MATHEMATICS

MATHEMATICS

The Numbers Behind the Neon

Mark Bollman

Albion College
Albion, Michigan, USA

CRC Press
Taylor & Francis Group
Boca Raton   London   New York

CRC Press is an imprint of the
Taylor & Francis Group, an **informa** business

A CHAPMAN & HALL BOOK

CRC Press
Taylor & Francis Group
6000 Broken Sound Parkway NW, Suite 300
Boca Raton, FL 33487-2742

Printed on acid-free paper
Version Date: 20140418

International Standard Book Number-13: 978-1-4822-0893-1 (Paperback)

Library of Congress Cataloging-in-Publication Data

Bollman, Mark, author.
    Basic gambling mathematics : the numbers behind the neon / Mark Bollman.
       pages cm
    Includes bibliographical references and index.
    ISBN 978-1-4822-0893-1 (paperback)
    1. Chance. 2. Probabilities. 3. Gambling--Mathematics. I. Title.

QA273.B596 2014
795.01'5192--dc23   2014013185

Visit the Taylor & Francis Web site at
http://www.taylorandfrancis.com

and the CRC Press Web site at
http://www.crcpress.com

For Laura,
who took a gamble with me in Las Vegas that's still paying off.

Contents

Preface

This book grew out of several years of teaching about gambling in a variety of contexts at Albion College beginning in 2002. For several years, I taught a first-year seminar called "Chance," which I came to describe as "probability and statistics for the educated citizen" as distinguished from a formula-heavy approach to elementary statistics. I also focused more on probability than statistics in Chance. Part of probability is gambling, of course, and so over the years, the course evolved to include more casino examples in class, whether by simulation or actual in-class game play. The course included a field trip to the Soaring Eagle Casino in Mount Pleasant, Michigan, late in the semester after all of the students had turned 18. This provided the students with a fine opportunity to combine theory with practice and see for themselves how the laws of probability worked, in a way that no classroom activity could mimic.

Later on, I expanded the gambling material into a course called Great Issues in Humanities: Perspectives on Gambling, in Albion's honors program. The course combined mathematics from Chance (for mathematics, in the words of one of my colleagues, is the first of the humanities) with other readings from literature, philosophy, and history to provide a well-rounded view of a subject that is not becoming less important in America.

Throughout my years teaching about gambling, I struggled to find a good probability textbook that covered the topics germane to my course without a lot of material that was not related to gambling. *Basic Gambling Mathematics* is my effort to distill the mathematics involved in gambling, and only that mathematics, into one place. While the final product started out as that textbook, over the course of an intensive summer spent writing (with a goal of 1000 words a day), it evolved to include more general information on the mathematics that I have found so fascinating during my years of teaching.

The text necessarily contains a large number of examples, illustrating the mathematical ideas in a range of casino games. The end of an example is indicated with a ■ symbol.

The exercises provided here are included for students, of course, and for those casual readers who would like to try their hand at some casino-related computations. Most of them either present other examples of the ideas in the chapter or ask for fairly straightforward verification of computations mentioned in the main text. Answers to all of the odd-numbered exercises are provided, and complete solutions to all exercises may be found in the separate solutions manual.

Acknowledgments

This book would not exist if not for my wife Laura, who encouraged me to write during the summer of 2012. I continue to be thankful for her support and encouragement as this project moved through the publication process to the finished work.

I would like to thank everyone at CRC Press/Taylor & Francis who have guided this project through to publication, including Sunil Nair, Sarah Gelson, Shashi Kumar, Robin Lloyd-Starkes, and the anonymous reviewers whose comments greatly enhanced the final version. It is genuinely impressive how much a look from a set of fresh eyes can enhance a project.

Thanks also to my colleagues, past and present, in Albion's Department of Mathematics and Computer Science for their ongoing collegiality and good spirits during challenging times. This book may not have been what the slot machine in my office was leading up to, but perhaps it explains, in part, why that distraction mattered.

Toward the end of the writing and revision process, I had the opportunity to teach a short course, Mathematics of the Gaming Industry, on these topics and so class-test parts of the text. The course included a 4-day trip to Las Vegas to compare theory and practice. My thanks go to the students in this course: Preston Arquette, Stefan Blachut, Cara Delaney, Sarah Erdman, Rebecca Guntz, Katie Strunk, Robin Todd, and Laura VerHulst, for their feedback on portions of chapters 2–7.

Spider craps (see page 73) was developed by Albion College student Jacob Engel during a summer research program in 2011. Funding for this project was provided by Albion's Foundation for Undergraduate Research, Scholarship, and Creative Activity (FURSCA).

Introduction

1.1   HISTORICAL BACKGROUND

The roots of probability lie in gambling. While the mathematical foundations of probability date back fewer than 400 years, evidence of games of chance can be found among the artifacts of far-older civilizations. Dice, for example, date back many thousands of years, in much the form we know today. Dungeons & Dragons players, who use polyhedral dice with 4–20 sides, may well notice that their dice bear Arabic numerals, in contrast to cubical casino and ordinary game dice, which use dots to label sides. There is a reason for this, as stated in [64]: Dice are older than numbers. Standard Arabic numerals only attained their final form around 700 CE, while cubical dice have been found dating back as far as 3000 BCE.

The transition from simple games of chance to a mathematical theory of probability really began in the 16th century when Girolamo Cardano wrote *Liber de Ludo Aleae* or *Book on Games of Chance*. "Aleae" refers here to games played with dice [5, p. 47]. The book was not published until 1663, nearly 100 years after Cardano's death—nonetheless, many of the ideas used to analyze casino games can be traced back to this work.

In this volume, Cardano gave the first mathematical treatment of *expectation*, which would come to be a fundamental idea in gambling mathematics. Looking back, we can also see that the notion of *sample space* is present, and that concept is also central to a meaningful mathematical treatment of gambling. By the end of the book, Cardano's work was showing the first signs of a theoretical, rather than experimental, approach to probability [5, p. 53].

Further progress in the theory of probability can be found in Galileo's treatise on the probabilities of rolling various numbers on three standard dice. In *Sopra le Scoperte Dei Dadi*, he correctly counted the various ways to roll sums such as 9, 10, 11, and 12, and in so doing eclipsed previous incorrect reasoning that had confounded dice players [24]. Galileo's work showed, through a simple enumeration of cases, that 10 is more likely than 9 and 11 more likely than 12.

The continued progress, after Cardano and Galileo, of probability as a formal mathematical subject can be traced to a 1654 letter from Antoine Gombaud, the Chevalier de Méré, to French mathematician Blaise Pascal. In this letter, Gombaud reported his experience at two different gambling games, and noted that his actual results were quite different from the results he expected based on his assessment of the probabilities. The goals of the two games he played were these:

1. To throw at least one 6 in four tosses of a fair six-sided die.

2. To throw at least one 12 (double sixes) in 24 tosses of two fair dice.

Gombaud's informal reasoning had led him to believe that his probability of winning either game was $\frac{2}{3}$, but he reportedly found that he won only slightly more often than he lost in the first game and lost slightly more often than not in the second. It is not hard to follow his reasoning: in game #1, he had 4 tries at a game with 6 possible outcomes, suggesting a $\frac{2}{3}$ chance of success; game #2 offered 24 shots at a game with 36 outcomes, leading to the same fraction.

It is also easy to find the flaw in this line of thought. Continuing game #1 for two more rolls, Gombaud would conclude that he would get at least one 6 every time he rolled a fair die six times, and it is not hard to imagine a case where this would not happen. Yahtzee and Settlers of Catan players, for example, are quite familiar with repeated inability to roll a desired number.

Pascal began a correspondence with Pierre de Fermat about these questions, which quickly grew to encompass related questions about games of chance. From these letters emerged a mathematical treatment of chance and uncertainty that laid the foundation for probability's development as a rigorous branch of mathematics. These principles would soon find application in a wide range of fields beyond gambling.

Over the ensuing years, the foundations of probability were refined; in the early 20th century, Andrei Kolmogorov stated a set of three axioms for probability that gave probability the same logical foundation as other branches of mathematics (see page 13). Beginning with Kolmogorov's three axioms, it is possible to derive the theory of probability in complete mathematical rigor.

The Rise Of Gambling In the USA

At about the same time, in 1931, Nevada paved the way for the spread of legal gambling in America by legalizing gambling within the state [64, p. 354–5]. Over the next few decades, gambling thrived in Reno, Las Vegas, and other Nevada cities, but there was no spread of legalized casino gambling to other states until 1976, when New Jersey voters approved a measure to allow casinos in Atlantic City. Casinos opened on that city's Boardwalk in 1978 [64].

The first Native American casino in the USA was launched in a garage in Zeba, Michigan, on December 31, 1983. The Pines faced legal challenges for 18 months before being forced to close [78]. However, the push for reservation casinos continued long after the first casino shut down. In 1988, the landscape of legalized gambling in the USA was changed forever when the Indian Gaming Regulatory Act issued guidelines for regulation of casinos run by Indian tribes on reservation lands. In the years since the act was passed, Native Americans have opened casinos across America, and voter referenda in numerous other states have paved the way for state-regulated casinos. Some sort of legal gambling, including casinos, dog and horse racing, and lotteries, is now available in 48 US states, all except Hawaii and Utah.

1.2 MATHEMATICAL BACKGROUND

Set Theory

We adopt a set-theoretic approach to probability in this text. In doing so, we assume that the meaning of the term *set* is understood, and so we will not define it explicitly. A deck of playing cards is a good illustration of the concept of a set. Considered one way, it's 52 separate things, but we may just as easily think of it as a single object.

Informally, we may reasonably think of a set as a collection of objects—called *elements*—considered as a unit. We will usually denote sets by capital letters and elements by lower-case letters. If A is a set and a is an element of A, we denote that relationship by $a \in A$. It is customary to use braces to enclose a list of the elements of a set.

Example 1.2.1. The set of *natural numbers* is $\{1, 2, 3, \ldots\}$ and is denoted by $\mathbb{N}$. $4 \in \mathbb{N}$, but $2.5 \notin \mathbb{N}$: 2.5 is not a natural number.

Example 1.2.2. The set of numbers that can appear when a standard six-sided die is rolled can be written as $A = \{1, 2, 3, 4, 5, 6\}$.

Example 1.2.3. A standard deck of playing cards consists of 52 cards, 13 cards in each of four suits. The suits are clubs (♣), diamonds (♢), hearts (and spades (♠). Clubs and spades are black; hearts and diamonds are red. Th 13 cards within each suit are denoted ace, 2, 3, 4, 5, 6, 7, 8, 9, 10, jack, queen, and king. Aces may, depending on the card game being played, be considered as either high or low.

We can think of a standard deck as a single set D containing 52 elements, which may be written systematically by listing the cards from 2 through ace in each suit and the four suits in succession, as

$$D = \{2\clubsuit, 3\clubsuit, \ldots, A\clubsuit, 2\diamondsuit, \ldots, A\diamondsuit, 2\heartsuit, \ldots, A\heartsuit, 2\spadesuit, \ldots, A\spadesuit\}.$$

It is possible, even desirable, to combine two sets in various ways to create new sets, or to consider a given set as arising from other sets by one of these operations. Two of the most common ways to combine sets are union and intersection.

Definition 1.2.1. The *union* of two sets A and B, denoted $A \cup B$, is the set of all elements belonging either to A or B, or both.

When using set-theoretic notation to describe a set described with a sentence, one indicator that you are dealing with a union may be the presence of the word "or."

Example 1.2.4. A standard deck of playing cards D may be thought of as the union of one set R of 26 red cards and another set B of 26 black cards. Symbolically, we would denote this union by $D = R \cup B$. A verbal description of the union might be "all cards that are either red or black."

Example 1.2.5. Two relatively high-ranking poker hands are a *straight* which consists of five cards in numerical sequence, and a *flush*, which is composed of five cards of the same suit. (A *straight flush* is a very rare hand consisting of five cards in sequence *and* of the same suit.) Suppose that you have been dealt the 3, 4, 5, and 6 of clubs from a standard deck. The set S all possible fifth cards that will complete a straight contains eight elements: all of the 2's and 7's. The set F of all possible cards that will complete a flush consists of the nine remaining clubs. Accordingly, the set of cards that will complete a straight *or* a flush is $S \cap F$. This union contains 15 cards: the 2 cards (2♣ and 7♣) that appear in both sets are only counted once each in the union.

Definition 1.2.2. The *intersection* of two sets A and B, denoted $A \cap B$ the set of all elements belonging to both A and B.

In translating between English and mathematics, intersections frequently correspond to the word "and."

Example 1.2.6. If A is the set of all clubs in a standard deck of cards and is the set of all 3s, then their intersection is the single card common to both sets: $A \cap B = \{3♣\}$. This card is both a 3 and a club.

Unions and intersections can be extended to more than two sets, as in the following example.

Example 1.2.7. We can think of a deck of cards as the union of four sets: clubs, diamonds, hearts, and spades. Using the first letter of each suit's name as an abbreviation for the set of cards of that suit, we could denote a standard 52-card deck by $C \cup D \cup H \cup S$.

Definition 1.2.3. We say that the set B is a *subset* of the set A if every element of B is also an element of A, and we write $B \subset A$.

Example 1.2.8. If we roll two standard dice and add the numbers that result, the set of all possible outcomes is $A = \{2, 3, 4, 5, 6, 7, 8, 9, 10, 11, 12\}$. In the casino game *craps*, the subset $B = \{7, 11\} \subset A$ is the set of rolls that is an automatic win for the shooter on the first roll.

It should be noted that, according to this definition, every set A is a subset of itself. If $B \subset A$ and we wish to exclude the possibility that $B = A$, we write $B \subsetneq A$. If $A \subset B$ and $B \subset A$ are both true, it follows that $A = B$.

An important set is the set with no elements, called the *empty set*. The empty set is denoted by $\emptyset$ or by $\{\}$. The empty set is a subset of every set. At the other extreme from $\emptyset$ is the *universal set* $\mathcal{U}$, which is the set that contains every element under consideration.

Definition 1.2.4. The *complement* of a set A is the set consisting of all elements that are in the universal set but not in A. We denote the complement of the set A by A'.

It is an immediate consequence of this definition that $A \cup A' = \mathcal{U}$ $A \cap A' = \emptyset$.

Example 1.2.9. Suppose again that we're drawing one card from a standard deck. Let $\mathcal{U}$ be the set of all 52 cards. If $A =$ the set of all red cards, then A the set of all cards that are not red—that is, the set of all black cards.

For our work in probability, we will frequently be interested in the size of a set—that is, how many elements it has. For convenience, we introduce the following notation: The expression $\#(A)$ denotes the number of elements in a set A.

This is most often used when A is a finite set—while it is certainly possible to consider the size of an infinite set, such sets are uncommon in gambling mathematics and are not considered in this book.

Example 1.2.10. If A is a standard deck of playing cards, then $\#(A$ 52.

Example 1.2.11. Suppose that we roll three standard dice: one each in red, green, and blue. Let the ordered triple (r, g, b) indicate the result of the roll in the order red, green, blue—so (2, 3, 4) is a different outcome from (3, 4, 2). Denote by A the set of all possible ordered triples resulting from one ro Each of r, g, and b is an integer in the range 1–6, and so we can write

$$A = \{(r, g, b) : 1 \leq r \leq 6, 1 \leq g \leq 6, 1 \leq b \leq 6\}.$$

It follows that $\#(A) = 6 \cdot 6 \cdot 6 = 216$.

The challenge here is that a set of items that are of interest in gambling mathematics can be very large. If we are interested in the set A of all possible five-card poker hands, then $\#(A) = 2,598,960$, and we'd like to have a way to come up with that number without having to list all of the hands and count them. Techniques for finding the size of such large sets will be discussed in Section 2.4.

Summation Notation

In much the same way that mathematicians have adopted shorthand notation for unions and intersections, there is an alternate notation that is useful for sums of many numbers. If we wish to add up a sequence of numbers numbers denoted by $x_1, x_2, \ldots, x_n$, we can express that by

$$\sum_{i=1}^{n} x_i,$$

where the Σ is the capital Greek letter sigma. Sigma corresponds to the letter S and is used here to stand for "sum." The index i is a new variable that is used to count the terms being added, and the subscript and superscript on the Σ indicate that the variable i starts at 1 and runs through n.

Example 1.2.12. The expression

$$\sum_{i=1}^{5} i$$

instructs us to compute the sum of the integers from 1 to 5, a sum which is, of course, 15.

Example 1.2.13.

$$\sum_{i=1}^{6} \frac{1}{i} = 1 + \frac{1}{2} + \frac{1}{3} + \frac{1}{4} + \frac{1}{5} + \frac{1}{6} = \frac{49}{20} = 2.45.$$

Example 1.2.14.

$$\sum_{i=1}^{10} i^2 = 1 + 4 + 9 + \ldots + 81 + 100 = 385.$$

Under certain circumstances, we can extend the upper limit of the sum to add up an infinite number of terms; this takes the form

$$\sum_{i=1}^{\infty} x_i$$

and the resulting sum is called an *infinite series*. The theory of infinite series, including the circumstances under which an infinite series *converges*, or has a finite sum, is a well-developed area of mathematics that is beyond the scope of this book. The infinite series that we encounter here will all converge.

The idea of indexing notation can also be applied to denote unions and intersections of more than two sets if n, the number of sets involved, is large. We may then use the alternate notation

$$\bigcup_{i=1}^{n} A_i$$

for the union of the n sets A_1 through A_n, which is often easier than $A_2 \cup A_3 \cup \cdots \cup A_n$. The corresponding notation for their intersection is

$$\bigcap_{i=1}^{n} A_i.$$

1.3   WHAT DOES IT MEAN TO BE RANDOM?

Throughout our consideration of probability applied to gambling, we will have occasion to address the notion of *randomness*. A formal definition of "random" inherently includes humans:

Definition 1.3.1. A process is *random* if its output contains no pattern that is detectable to human observers.

This is not as satisfying a definition as we might want, but for practical purposes, it does a pretty good job of explaining what we mean when we describe something as "random." An alternate approach to randomness is this: A process is random if every possibility or every arrangement of its components is equally likely.

In a modern casino, the slot machines and video poker machines are controlled by computer chips that generate thousands of numbers every second. Those numbers are determined by a complicated algorithm that we could, in theory, exploit to predict the exact outcome of each spin of the (real or simulated) reels. Since an algorithm generates these numbers, they are properly termed *pseudorandom* numbers: not strictly random in a technical sense, but random enough for their intended use.

Example 1.3.1. The TI-58C calculator manufactured by Texas Instruments in the 1970s included a random number generator that used the following algorithm to generate a list of pseudorandom numbers [42, p. 54]:

1. Enter an initial number x_0 in the range $0 \leq x_0 \leq 199,017$.

2. Define
$$x_{n+1} = (24,298 \cdot x_n + 99,991) \mod 199,017,$$

   where "mod 199,017" denotes the remainder when $24,298 \cdot x_n + 99$, is divided by 199,017.

3. Scale x_{n+1} as necessary to produce a random integer in a desired range. If the range is from 1 to B, the equation

$$y_{n+1} = (x_{n+1} \mod B) + 1$$

will produce an integer in that interval.

Step 2 of this algorithm will generate a list of integers in the range 0–199,016. If we begin with $x_0 = 1146$, for example, we get the sequence

$$x_1 = 83,119, x_2 = 100,937, x_3 = 180,726, x_4 = 70,234, x_5 = 74,948.$$

If we then scale these integers to the range 1–6, to represent rolls of a standard die, we would get 2, 6, 1, 5, 3.

Some attempts to cheat casinos at keno or video poker have relied on insider knowledge of these algorithms. In practice, however, doing so is so difficult that the machines are "random enough" for their intended purpose. It may well be the case that the numbers generated by the computer chip repeat with a period of 32 million, but it is equally the case that no person can hope to take advantage of this repetition. Contemporary slot machines generate numbers at a rapid rate even when they're not being played; it is only the pull of a lever or the push of a button that translates the generated number into a sequence of symbols and generates the corresponding payoff. Changing the time of the triggering event by as little as one millisecond will result in a different outcome—this is too fast for humans to exploit.

Example 1.3.2. A standard deck of playing cards is arranged in a specified order at the factory, frequently all of the clubs in order from king down to ace, followed by the diamonds, hearts, and spades. This is about as far from random (in the sense of Definition 1.3.1) as the deck can get, and so it is important that a new deck be thoroughly shuffled before being put in play. How thorough is "thoroughly shuffled"? Seven standard riffle shuffles, according to Dave Bayer and Persi Diaconis—provided that the shuffles are "imperfect" [36]. Perfect shuffles are those where the deck is divided exactly into two 26-card halves and the cards from each half are perfectly alternated when interlaced. These have the curious property that if the top card of the original deck remains on top throughout the process, eight perfect shuffles return the deck to its original non-random state [84].

Bayer and Diaconis showed that fewer than seven shuffles were not sufficient to randomize the deck, and that more than seven didn't improve the randomness significantly.

Example 1.3.3. By contrast, the street game of *Three-Card Monte* is anything but random. The setup is simple enough: a dealer flips three face-down playing cards, two red and one black, back and forth rapidly, and challenges the player to choose the black card. The machinations used by the dealer to mix the cards can be incredibly intricate.

The game could not be more simple to play: put down your money and try to pick the right card out of three. A one-in-three chance of winning is better than you get from many traditional casino games or lotteries, and if you can get just a little bit lucky, or find some kind of telltale pattern in the dealer's routine, Three-Card Monte might seem like a reasonable gamble.

It's not.

The dealer has complete control over the placement of the cards, and a skilled Three-Card Monte dealer, or *grifter*, will know where the odd card is at all times, and will control the game environment so that every outside gambler loses. Indeed, any player you see win at Three-Card Monte is a *shill*—someone who is working in league with the dealer to take all the players' money and has been tipped off to the location of the black card by the dealer's words or actions. A thorough discussion of the methods used by Three-Card Monte grifters may be found in [96].

1.4 EXERCISES

Answers begin on page 253.

An American roulette wheel contains 38 pockets, one for each number from 1 to 36, one numbered 0, and one numbered 00. The 0 and 00 pockets are green; the numbers from 1 to 36 are colored either red or black, as follows:

Red numbers: 1,3,5,7,9,12,14,16,18,19,21,23,25,27,30,32,34,36.
Black numbers: 2,4,6,8,10,11,13,15,17,20,22,24,26,28,29,31,33,35.

See pages 20 and 21 for images of a roulette wheel and the betting layout.

Roulette numbers may also be classified as low (1–18) or high (19–36), and as odd or even. Note that 0 and 00 are neither even nor low. If we denote the sets of red, black, high, low, odd, and even numbers by R, B, H, L, O, and E respectively, find the elements of the following sets and describe each in words.

1.1. $R \cap O$

1.2. $B \cup E$

1.3. $R \cap H \cap E$

1.4. R'

1.5. $L \cap H$

1.6. $(L \cup O \cup B)'$

1.7. $(H \cup R)'$

1.8. Using the random number generator in Example 1.3.1, find the first six random numbers generated with an initial seed of $x_0 = 48,101$ and scale these numbers to the range 1–8 to simulate six successive rolls of an eight-sided die.

Hint: This process can be streamlined on a TI-84+ calculator by using the $\boxed{\text{Ans}}$ key. This key inserts the answer of the previous calculation into a new calculation. Once the first result has been found, the sequence

$$24{,}298 \;\boxed{\times}\; \boxed{\text{Ans}} \;\boxed{+}\; 99{,}991 \;\boxed{\text{ENTER}}$$

computes the next number in the sequence. Repeatedly pressing $\boxed{\text{ENTER}}$ will re-enter that expression on the command line and evaluate it.

It is still necessary to handle the "mod 199,017" part of the RNG, of course. The **remainder** function will do that; it can be found by pressing $\boxed{\text{MATH}}$ scrolling right to the **NUM** menu, and choosing option 0. To compute the remainder when x is divided by y, enter **remainder(x,y)**.

Fundamental Ideas

2.1 DEFINITIONS

We begin our study of probability with the careful definition of some important terms.

Definition 2.1.1. An *experiment* is a process whose outcome is determined by chance.

This may not seem like a useful definition. We illustrate the concept with several examples.

Example 2.1.1. Roll a standard six-sided die and record the number that results.

Example 2.1.2. Roll two standard six-sided dice (abbreviated as 2d6) and record the sum.

Example 2.1.3. Draw one card from a standard deck and record its suit.

Example 2.1.4. Roll 2d6 and record the larger of the two numbers rolled (or the number rolled, if both dice show the same number).

Example 2.1.5. Deal a five-card video poker hand and record the number of aces it contains.

An important trait of an experiment is that it leads to a definite outcome. While we will eventually concern ourselves with individual outcomes, we begin by looking at all of the possible results of an experiment.

Definition 2.1.2. The *sample space* **S** of an experiment is the set of all possible outcomes of the experiment.

Example 2.1.6. If we consider the simple experiment of tossing a fair coin, our sample space is **S** = {Heads, Tails} or {H,T} for short.

Example 2.1.7. In Example 2.1.1, the sample space is $\mathbf{S} = \{1, 2, 3, 4, 5,$
The same sample space applies to the experiment described in Example 2.1.4.

Example 2.1.8. In Example 2.1.2, the sample space is

$$\mathbf{S} = \{2, 3, 4, 5, 6, 7, 8, 9, 10, 11, 12\}.$$

It is important to note that the 11 elements of $\mathbf{S}$ in this example are not equally likely, as this will play an important part in our explorations of probability. Cardano himself recognized that the proper sample space (or "circuit," to use his term) for questions involving the sum of two dice has 36, not 11, elements.

Example 2.1.9. $\mathbf{S} = \{\clubsuit, \diamondsuit, \heartsuit, \spadesuit\}$ is the sample space in Example 2.1.3.

When we're only interested in some of the possible outcomes of an experiment, we are looking at subsets of $\mathbf{S}$. These are called *events*.

Definition 2.1.3. An *event* A is any subset of the sample space $\mathbf{S}$. An event is called *simple* if it contains only one element.

Example 2.1.10. In Example 2.1.2, the event $A = \{$The roll is an even #
can be written $A = \{2, 4, 6, 8, 10, 12\}$. The event $B = \{$The roll is a square
is $B = \{4, 9\}$. The event $C = \{$The roll is a prime number greater than 10
is $C = \{11\}$ and is a simple event—though this is certainly not the simples
verbal description of C.

Example 2.1.11. In playing video poker, suppose that your initial dealt hand
is $4\heartsuit 5\heartsuit 6\heartsuit 7\heartsuit J\clubsuit$ and you discard the $J\clubsuit$ to draw a new fifth card in hopes of
completing either a straight or a flush, the sample space for the draw contains
the 47 remaining cards. The event corresponding to succeeding at this goal is
the set A consisting of 15 cards: the 9 remaining hearts, the 3 nonheart 3s (we
exclude the $3\heartsuit$ here because we already included it among the hearts), and
the 3 nonheart 8s.

Definition 2.1.4. Two events A and B are *disjoint* if they have no elements
in common—that is, if $A \cap B = \emptyset$.

Example 2.1.12. In Example 2.1.3, where we draw one card from a standard
52-card deck and observe its suit, the two events $D = \{$The card is a $\diamondsuit\}$
$H = \{$The card is a $\heartsuit\}$ are disjoint, since no card has more than one suit.

2.2 AXIOMS OF PROBABILITY

In any formal mathematical system, it is necessary to specify certain statements, called *axioms* or *postulates*, which are assumed to be true without the need for proof. We may think of our axioms as the foundation on which our mathematical system is constructed, with each theorem resting atop results that are required in its proof. Ideally, these statements should be small in number; if there are too many axioms, proofs tend to be trivial—too many theorems are true by assumption. Additionally, axioms should be results that are "obvious" to reasonable observers.

Euclidean geometry was the first mathematical system founded on this axiomatic method. In Euclid's original formulation, five "common notions"— statements such as "If $a = b$ and $a = c$, then $b = c$," assumed to be true throughout all of mathematics—were joined by five postulates of a more specifically geometric nature to form the logical foundation from which all other geometric results followed. The fifth postulate ("If two lines are cut by a transversal in such a way that the interior angles on one side of the transversal add up to less than two right angles, then the lines intersect on that side of the transversal,") was considerably less intuitively obvious than the first four, and so mathematicians over the next several centuries devoted considerable effort to trying to prove it from the other axioms, without success.

The following three axioms, which were precisely formulated in 1933 by the Russian mathematician Andrei Kolmogorov, will form the foundation of our work in probability.

Axiom 1. *Given a sample space S, it is possible to assign to each event E nonnegative number $P(E)$, called the **probability** of E.*

If an event is certain to occur, then its probability is 1. If an event is impossible, then its probability is 0.

Axiom 2. $P(S) = 1$.

In a given experiment, something must happen.

Axiom 3. *If $A_1, A_2, \ldots, A_n$ are pairwise disjoint events, then*

$$P\left(\bigcup_{i=1}^{n} A_i\right) = \sum_{i=1}^{n} P(A_i).$$

In words, Axiom 3 states that the probability of a union of disjoint events is equal to the sum of the probabilities of the individual events. It should be noted that the events under consideration must be disjoint—if they are not, a slightly more complicated formula called the *second addition rule* may be used to find the probability of their union. We shall consider this rule in Chapter 3.

As an example of how these axioms can be used to construct proofs, Axioms 2 and 3 together allow us to prove the following (fairly obvious) result:

Theorem 2.2.1. $P(\emptyset) = 0$.

Proof. Since $\emptyset$ has no elements, it follows that $\emptyset$ and $\mathbf{S}$ have no elements in common—thus they are disjoint. By Axiom 3, we have

$$P(\emptyset \cup \mathbf{S}) = P(\emptyset) + P(\mathbf{S}).$$

Since $\emptyset \cup A = A$ for any set A, we have $\emptyset \cup \mathbf{S} = \mathbf{S}$, and so it follows that

$$P(\mathbf{S}) = P(\emptyset) + P(\mathbf{S})$$

or, applying Axiom 2,

$$1 = P(\emptyset) + 1,$$

from which the conclusion immediately follows by subtraction.

More generally, and more usefully, we have the following result, called the *complement rule*:

Theorem 2.2.2. *(The Complement Rule)*: *For any event A,*

$$P(A') = 1 - P(A).$$

Proof. Given any event A, we know that $A \cap A' = \emptyset$ and $A \cup A' = \mathbf{S}$, so that A and A' are disjoint sets whose union is the entire sample space. It follows from Axiom 3 that

$$P(A \cup A') = P(A) + P(A').$$

We also have

$$P(A \cup A') = P(\mathbf{S}) = 1,$$

using Axiom 2. Combining these two equations gives

$$P(A) + P(A') = P(\mathbf{S}),$$

hence

$$P(A') = P(\mathbf{S}) - P(A) = 1 - P(A),$$

as desired.

We shall see that the complement rule frequently turns out to be useful in simplifying probability calculations. It can, for example, be used to resolve the Chevalier de Méré's question that launched probability theory as a branch of mathematics (see page 54).

A more significant result limits probabilities to the interval $0 \le P(E) \le$

Theorem 2.2.3. *For any event E, $0 \le P(E) \le 1$.*

Proof. We note that $E \cup E' = \mathbf{S}$, and furthermore that $E \cap E' = \emptyset$. By Axiom 3,

$$P(E \cup E') = P(E) + P(E')$$

and by Axiom 2,

$$P(E \cup E') = P(\mathbf{S}) = 1.$$

Combining these two equations gives

$$P(E) + P(E') = 1.$$

Axiom 1 tells us that probabilities are nonnegative, so $P(E) \geq 0$ and half of our conclusion is established. If $P(E) > 1$, it follows that $P(E') = 1$ $P(E) < 0$, an impossibility. This contradiction establishes the other half of our conclusion: $P(E) \leq 1$, completing the proof.

We could have chosen to use Theorem 2.2.3 in place of or in addition to Axiom 1, but since we can prove the former from the latter, and since Axiom 1 is simpler, it is the preferred choice as an axiom. It is a worthy goal, in constructing an axiomatic system, not to include anything in an axiom that can be proved from the other axioms—we say then that our axioms are *independent* of each other.

To progress further, we need to develop procedures for assigning probabilities to events. We define $P(A)$ as a ratio of the size of A to the size of the sample space $\mathbf{S}$:

Definition 2.2.1. Let $\mathbf{S}$ be a sample space in which all of the outcomes are equally likely, and suppose $A \subset \mathbf{S}$. The *probability* of the event A is

$$P(A) = \frac{\text{Number of elements in } A}{\text{Number of elements in } \mathbf{S}} = \frac{\#(A)}{\#(\mathbf{S})}.$$

For convenience, we state this as a definition, although it is possible to prove this formula from Kolmogorov's axioms. The following simple theorem follows immediately from this definition:

Theorem 2.2.4. *If $B \subset A$, then $P(B) \leq P(A)$.*

Proof. If $B \subset A$, the definition of a subset tells us that every element of is also an element of A, but not necessarily vice versa: that is, A contains at least as many elements as B. Accordingly:

$$\#(B) \leq \#(A).$$

Dividing by $\#(\mathbf{S})$ gives

$$\frac{\#B}{\#(\mathbf{S})} \leq \frac{\#A}{\#(\mathbf{S})},$$

or

$$P(B) \leq P(A),$$

as desired.

There are several ways by which we might determine the value of $P($ These methods vary in their mathematical complexity as well as in their level of precision. Each of them corresponds to a question we might ask or try to answer about a given probabilistic situation.

1. **Theoretical Probability**

   If we are asking the question *"What's supposed to happen?"* and relying on pure mathematical reasoning rather than on accumulated data, then we are computing the *theoretical probability* of an event.

   Example 2.2.1. Consider the experiment of tossing a fair coin. Since there are two possible outcomes, heads and tails, we can compute the theoretical probability of heads as $1/2$.

   Example 2.2.2. If we roll 2d6, what is the probability of getting a sum of 7?

   An incorrect approach to this problem is to note that the sample space is $\mathbf{S} = \{2, 3, 4, 5, 6, 7, 8, 9, 10, 11, 12\}$, and since one of those 11 outcomes is 7, the probability must be $\frac{1}{11}$. This fails to take into account the fact that some rolls occur more frequently than others—for example, while there is only one way to roll a 2, there are 6 ways to roll a 7: 1-6, 2-5, 3-4, 4-3, 5-1, and 6-1. (It may be useful to think of the dice as being different colors, so that 3-4 is a different roll from 4-3, even though the numbers showing are the same.) Counting up all of the possibilities shows that there are $6 \cdot 6 = 36$ ways for two dice to land. Since six of those yield a sum of 7, the correct answer is $P(7) = \frac{6}{36} = \frac{1}{6}$.

2. **Experimental Probability**

   When our probability calculations are based on actual experimental results, the resulting value is the *experimental probability* of A. Here, we are answering the question *"What really did happen?"*

   Example 2.2.3. Suppose that you toss a coin 100 times and that the result of this experiment is 48 heads and 52 tails. The experimental probability of heads in this experiment is $48/100 = .48$, and the experimental probability of tails is $52/100 = .52$.

   This experimental probability is different from the theoretical probability of getting heads on a single toss, which is $\frac{1}{2}$. This is not unusual.

   One area of gambling where experimental probability is especially prevalent is blackjack. As we will see in Chapter 6, the mathematics of blackjack can be fairly complicated, owing to the wide variety of rules that are used in various casinos and the dependence of each hand on the hands previously dealt. Many blackjack probabilities have been computed only through computer simulation of millions of hands under specified game conditions.

3. Subjective Probability

Sometimes our probability calculations are intended to answer the question *"What happened in the past?"* and use the answer to that question to predict the likelihood of a future event. A probability so calculated is called the *subjective probability* of A. The question we're trying to answer here is about an event that has not yet happened, instead of one that has already occurred and for which we have generated data—in this way, subjective probability can be distinguished from experimental probability.

Example 2.2.4. Perhaps the best example of subjective probability is weather forecasting. It's simply not possible to measure such parameters as temperature, humidity, barometric pressure, and wind speed and direction, plug those numbers into a (possibly) complicated formula, and have a computer generate accurate long-range weather forecasts. The inherent instability of the equations that govern weather means that small differences in the input values can lead to large differences in the predicted weather. Accordingly, there will always be some uncertainty in weather predictions.

Las Vegas-based comedian George Wallace described the imprecision of weather forecasting in the following way:

> *[The weatherman] has the only job in the world where he's never right and they say "Come back tomorrow."...If you go to school, and the teacher asks you "How much is 3+4?" you can't go "Nearly 7, with a 20% chance of being 8, and maybe 9 in the low-lying areas."*

As subjective probability is, by its very nature, less mathematically precise than the other two types, we shall not consider it any further here.

The connection between theoretical and experimental probability is described in a mathematical result called the *Law of Large Numbers*, or LLN for short.

Theorem 2.2.5. *(Law of Large Numbers) Suppose an event has theoretical probability p. If x is the number of times that the event occurs in a sequence of n trials, then as the number of trials n increases, the experimental probability x/n approaches p.*

Informally, the LLN states that, in the long run, things happen in an experiment the way that theory says that they do. What is meant by "in the long run" is not a fixed number of trials, but will vary depending on the experiment. For some experiments, $n = 500$ may be a large number, but for others—particularly if the probability of success or failure is small—it may take far more trials before the experimental probabilities get acceptably close to the theoretical probabilities.

Example 2.2.5. While it is possible to load a die so that the sides are not equally likely (see Example 4.1.6 for one way), it is not possible to load a coin in that way—no matter what sort of modifications are made, short of putting the same face on both sides of a coin, the probability of heads and the probability of tails will always equal $\frac{1}{2}$ when the coin is tossed [21]. Nonetheless, it is customary when talking about tossing a coin to specify that the coin is "fair," and we shall do so here. Consider the experiment of repeatedly flipping a fair coin, where the theoretical probability of heads is $\frac{1}{2}$. We don't expect that every other toss will yield heads, but we do expect that the coin will land heads up approximately half the time. Table 2.1 contains data, and the corresponding experimental probabilities, from a computer simulation of a fair coin.

TABLE 2.1: Coin tossing and the Law of Large Numbers

# of tosses, n	# of Heads, x	P(Heads), $\frac{x}{n}$	P(Heads) - $\frac{1}{2}$
10	3	.30	-.20
100	48	.48	-.02
1000	518	.518	.018
10,000	5044	.5044	.0044
100,000	50,039	.50039	.00039
1,000,000	499,740	.499740	-.000260
10,000,000	4,999,909	.4999909	-.0000091

While the *number* of heads tossed (column 2) is not exactly half, we observe that, as the number of trials increases, the *proportion* of tosses landing heads (column 3) gets closer to $\frac{1}{2}$, and the value in the fourth column approaches 0, as the LLN suggests.

2.3 ELEMENTARY COUNTING ARGUMENTS

Whether we are computing theoretical or experimental probabilities, we will find ourselves counting things. For simple games, it is not always necessary to invoke complicated formulas; all we need to do is be careful to make sure that every element in the set at issue is counted once and no element is counted more than once.

Example 2.3.1. A game occasionally seen in carnivals, and also present for a time on a carnival midway at the Excalibur Casino in Las Vegas, has a *barker*, or carnival employee, offering to guess your birth month. The barker is considered correct if his or her guess is within two months of your actual birthday, so if you were born in August, the barker wins on a guess of June, July, August, September, or October. You win a prize if the barker's guess is any other month. The barker typically writes down a guess before asking for your birthday, thus eliminating any opportunity for either side to profit by lying.

No matter when you were born, there are five months that, if chosen by the barker, result in a loss for you, and seven on which you win. It follows immediately that your probability of winning is $\frac{7}{12}$, and your probability of losing is $\frac{5}{12}$.

While the barker can guess any month at random without affecting these probabilities, a smart strategy for the barker would look at the number of days covered by a prediction rather than the number of months, which is fixed at five. Assuming that all 365 birthdays (omitting February 29) are equally likely, the barker should choose a month to avoid catching February. A guess of any month from May through November covers three months with 31 days and two with 30, a total of 153 days. This is more days than can be covered by guessing any of the other five months, and so a barker should pick from those months exclusively. Of course, this strategy leaves the carnival open to an attack by hordes of people with February birthdays, as none of these seven months cover February in their five-month intervals.

The Big Six Wheel

Example 2.3.2. The *Big Six* wheel is a large wheel (usually six feet or more in diameter; the wheel at Bob Stupak's Vegas World measured 30 feet across) divided into 54 sectors. Each sector bears some kind of symbol, usually a piece of currency or casino logo. These wheels are frequently placed near the perimeter of the casino floor in full view of an entrance, as if to beckon a prospective gambler with the lure of a fun and simple game.

Fun? If you're winning, probably. Every game is fun when you're winning. Simple? Yes. Players bet on one of the wheel's symbols. The wheel is spun, and if the symbol at the top of the wheel, indicated by an arrow or a leather strip called a *clapper*, matches the one the player bets on, the bet wins.

The Big Six wheel draws its name from the number of different payoffs that are possible when it's spun. A standard wheel is divided into 54 sectors, and the following symbols appear in the indicated quantities:

Symbol	Count
$1	24
$2	15
$5	7
$10	4
$20	2
Logo A	1
Logo B	1

At the Venetian Casino in Las Vegas, the two logos on the wheel are a joker and an American flag. Down the Strip at the Cosmopolitan, the logo slots are labeled "Art" and "Music."

Calculating probabilities on the Big Six wheel is a simple matter of counting the sectors containing the symbol on which you have bet and dividing by 54. Accordingly, we have the following probabilities:

Symbol	Probability
$1	24/54
$2	15/54
$5	7/54
$10	4/54
$20	2/54
Logo A	1/54
Logo B	1/54

Roulette

> *My lucky number is 4 billion. That doesn't come in real handy when you're gambling.*

> —Mitch Hedberg, *Strategic Grill Locations*

In the game of *roulette*, a small ball is spun on a rotating wheel and comes to rest in a numbered pocket on the wheel. *European* roulette uses a wheel divided into 37 pockets, numbered 1 to 36 and 0. The 0 is colored green, and the remaining 36 slots are split evenly between red and black. In *American* roulette, a 38th pocket, colored green and numbered 00, is added—see Figure 2.1.

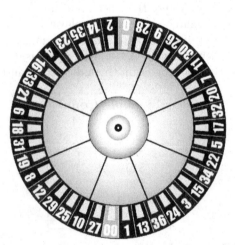

FIGURE 2.1: American roulette wheel [18]

Players select numbers from the wheel and are paid if the ball lands in a pocket bearing one of their numbers. A variety of bets, at a range of payoffs, is available to players—see Figure 2.2 for an illustration of the betting layout and Table 2.2 for a list of payoffs.

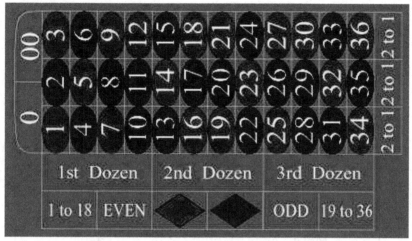

FIGURE 2.2: American roulette layout [17]

TABLE 2.2: Roulette bets and payoffs

Bet	# of numbers	Payoff
Straight	1	35 to 1
Split	2	17 to 1
Street	3	11 to 1
Corner	4	8 to 1
Basket[a]	5	6 to 1
Double street	6	5 to 1
Dozen[b]	12	2 to 1
Even-money[c]	18	1 to 1

Notes:

a—The *basket* bet is available only on American roulette wheels and can only be made on the five-number combination 0, 00, 1, 2, and 3.

b—The *dozen* bet may be made on the numbers from 1 to 12, 13 to 24, or 25 to 36, or on any of the three columns on the betting layout. These columns are depicted as rows in Figure 2.2.

c—An *even-money* bet may be made on odd, even, red, black, low (1–18), or high (19–36) numbers. (0 and 00 are neither even nor low.)

As a convenient shortcut, it should be noted that, except for the basket bet, the payoff for a bet on n numbers is $\dfrac{36-n}{n}$ to 1. A winning player's initial bet is returned with the payoff.

Example 2.3.3. The probability of winning a roulette bet is easy to compute:

$$P(\text{Win}) = \frac{\#\text{ of numbers covered by the bet}}{\#\text{ of numbers on the wheel}}.$$

In European roulette, the denominator is 37; in American roulette, it is 38. This formula produces the probabilities recorded in Table 2.3.

TABLE 2.3: Roulette probabilities

Bet	P(Win): European	P(Win): American
Straight	1/37	1/38
Split	2/37	2/38
Street	3/37	3/38
Corner	4/37	4/38
Basket	—	5/38
Double street	6/37	6/38
Dozen	12/37	12/38
Even-money	18/37	18/38

We can easily see that European roulette gives slightly higher probabilities of winning the same bet, so if you have a choice between European and American roulette, pick European. It's better for the player, though as we shall see, both games still favor the casino.

California state law forbids the use of a roulette wheel as the sole determiner of the outcome of a game of chance. Nonetheless, casinos in San Diego County have found a number of innovative ways to use a roulette wheel or other gaming devices to run a game that is mathematically equivalent. At the Barona Casino, a single-zero wheel is spun and 3 cards from a 37-card deck, numbered 0 to 36, are dealt to the table. If the number spun is in the range 0–12, the first card is flipped over and used as the result of the game. Similarly, the second card is used if the wheel comes up 13–24 and the third if the number on the wheel is 25–36. Harrah's simply uses a bingo or keno cage containing 76 balls. Two bear each number from 00 to 36 appearing on an American wheel, and drawing a ball substitutes for spinning a wheel [72].

Blackjack

Blackjack is one of the most popular casino games, in part because the gambler plays an active role in the game by his or her choices of how to play the cards that are dealt. The goal in blackjack is to get a total closer to 21 than the dealer, without going over 21. Aces count either 1 or 11, at the player's discretion. Face cards count 10, and all other cards count their rank. A full examination of blackjack may be found in Chapter 6; we examine here some simple questions that can be answered with the mathematics we have developed so far.

Example 2.3.4. Blackjack hands start with two cards dealt to each player. In a blackjack game dealt from a double deck—two standard decks shuffled together—find the probability that your first card is an ace.

There are 8 aces in a double deck, and 104 cards in all, so

$$P(\text{Ace}) = \frac{8}{104} = \frac{1}{13}.$$

Example 2.3.5. In a double-deck game, suppose that you have been dealt two 10s, for a total of 20. The dealer's first card is an ace. What is the probability that his second card will be a 10-count card (10, jack, queen, or king) that brings his total to 21 and beats you?

There are 101 cards as yet unaccounted for. Each deck starts out with 16 ten-count cards, for a total of 32 at the hand's start, and two of them are in your hand and so cannot be drawn by the dealer. This leaves 30 in the deck, and we have

$$P(\text{Dealer 21}) = \frac{30}{101} \approx .2970,$$

or slightly less than 30%.

Poker

Five-card poker hands are ranked in the order given in Table 2.4, from highest to lowest—or from least probable to most probable. It should be noted that suits have no relative rank in poker: a royal flush in spades is no better or worse than a royal flush in any other suit. This is in contrast to contract bridge, where during the bidding phase of the game, the four suits are ranked clubs, diamonds, hearts, and spades from lowest to highest.

TABLE 2.4: Five-card poker hands

Hand	Description
Royal flush	AKQJ10 of the same suit
Straight flush	5 cards of the same suit and in sequence
Four of a kind	Four cards of the same rank
Full house	3 of a kind and a pair
Flush	5 cards of the same suit, not in order
Straight	5 cards in consecutive ranks, not all the same suit
Three of a kind	3 cards of the same rank
Two pairs	Two pairs of cards of different ranks
One pair	Two cards of the same rank
High card	Five cards of different ranks and not all the same suit

These relative hand rankings are used in any form of poker using a standard deck. If wild cards are added to the deck, as when using a 53-card deck including a joker, some slight variations in hand frequency are introduced (see page 167). There are numerous variations on poker's fundamental idea, which is to achieve as high-ranking a hand as possible within the rules of the game at play.

Texas Hold'em is a form of poker whose popularity skyrocketed in the early 2000s, in large part due to its presence on television. The game is played as follows:

- Two cards—the *hole* or *pocket* cards—are dealt face down to each player. This is followed by a round of betting.

- Three cards—the *flop*—are dealt face up to the center of the table. These three cards and the two that follow are *community* cards, and are used by all players, together with their two hole cards, to make the best possible five-card poker hand. A second round of betting follows the flop.

- A fourth community card—the *turn*—is dealt face up to the table, and a third round of betting follows.

- The final community card—the *river*—is dealt to the table, completing the hands. A final round of betting takes place, and at the end, player with the highest five-card poker hand, made from the community cards and their two hole cards, remaining in the game wins the pot.

The fall of the cards at Texas Hold'em suggests many probability questions, some of which can be answered using no more than simple counting. Televised poker, whose popularity may be traced in no small part to the development of tiny cameras that allow home viewers to see the players' hole cards, usually includes the probability of each player winning the hand, which is calculated by a brute-force computer simulation. All possible choices of the cards remaining to be dealt are considered, and the winning hand for each arrangement is determined. Simple counting of the number of times each player wins gives the experimental probabilities shown on the screen.

Example 2.3.6. Suppose that your hole cards are the 4♣ and 5♠, the flop is 6♦ 2♡ 5♦, and the turn is the 6♡. What is the probability that the river will complete a full house?

To pull a full house, the river must be one of the two remaining 5s or the two 6s. Since there are 4 such cards, while 46 cards remain in the deck, you have probability $4/46 \approx .0870$ of drawing into a full house.

We have carried out this calculation without concern for what cards other players may hold and thus are unavailable to be drawn on the river. If all of the 5s and 6s have been dealt out, then your probability of completing your full house is 0, but if none of them is in other players' hands, then your chances have increased past .0870. In light of the information we have, .0870 is the best estimate for P(full house), and if we were to compute the probability for every possible game situation, we would find that the *average* probability of a full house would be .0870.

Example 2.3.7. In the example above, what is the probability that you will complete a straight on the river?

To fill in your straight, you need to draw a 3. Since there are four 3s as yet unaccounted for, the probability of getting a straight is also $4/46$.

There is considerable literature on Texas Hold'em that takes the game's mathematics into account. One of the best books is *Introduction to Probability with Texas Hold'Em Examples* by Frederic Paik Schoenberg [62]. A good portion of the game's strategy is when to make a bet or fold a given hand, and most experts would agree that the hand above should be folded before the flop, as two low cards, even consecutive ones, of different suits comprise a very weak starting hand.

Example 2.3.8. Find the probability that your hole cards are the same suit—"suited," in the game's lingo.

The first card doesn't matter; what we need to compute is the probability that your second card matches its suit. Again, with no knowledge about the other players' holdings, there are 12 cards of your suit remaining out of 51 overall, so the probability of suited hole cards is $12/51 = 4/17 \approx .2105$.

2.4 ADVANCED COUNTING ARGUMENTS

In many applications of Definition 2.2.1, $P(A) = \#(A)/\#(S)$, the sheer number of elements comprising an event or the sample space is far too large to be counted one by one. When computing probabilities, we seldom have need to consider each of these simple events individually; we are usually only interested in how many there are.

2.4.1   Fundamental Counting Principle

Combinatorics is the branch of mathematics that studies counting techniques. Frequently in gambling mathematics, we find ourselves considering the number of ways in which several events can happen in sequence. If we know the number of ways that each individual event can happen, elementary combinatorics tells us that simple multiplication can be used to find the answer.

Theorem 2.4.1. *(The Fundamental Counting Principle) If there are n independent tasks to be performed, such that task T_1 can be performed in m_1 ways, task T_2 can be performed in m_2 ways, and so on, then the number of ways in which all n tasks can be performed successively is*

$$N = m_1 \cdot m_2 \cdot \ldots \cdot m_n.$$

That the FCP is a reasonable result can be easily seen by testing out some examples with small numbers and listing all possibilities—for example, when rolling 2d6, one red and one green, each die is independent of the other and can land in any of six ways. By the FCP, there are $6 \cdot 6 = 36$ ways for the two dice to fall, and this may be confirmed by writing out all of the possibilities.

Proof. We shall prove the FCP using a technique called *mathematical induction*, which is described in Appendix B for those readers interested in the details of the proof. Let n represent the number of tasks to be performed. Assume at the outset that task T_i can be performed in m_i ways.

If $n = 1$, we are asked to verify that task T_1 can be performed in m_1 ways, which is true. (The product on the right side of the equation above contains only one factor.)

We assume next that the statement is true for $n = k$, that is, that tasks T_1 through T_k can be performed in

$$m_1 \cdot m_2 \cdot \ldots \cdot m_k$$

ways.

We now seek to prove that the statement is true for $n = k+1$. To do so, we divide the tasks T_1 through T_{k+1} into two groups, with T_1 through T_k in one group and T_{k+1} by itself in the second group. By the induction hypothesis, the tasks in the first group can be completed in $m_1 \cdot m_2 \cdot \ldots \cdot m_k$ ways. We can think of this succession of tasks as a single task, which we follow by performing task T_{k+1}. These two tasks can be performed in succession in

$$(m_1 \cdot m_2 \cdot \ldots \cdot m_k) \cdot m_{k+1}$$

ways, which is the number of ways to perform all $k + 1$ tasks in succession, completing the proof.

Example 2.4.1. Leaving out suits for the moment (since suits have no standing in poker) and considering only the ranks of the cards, how many possible pairs of hole cards are there in Texas Hold'em?

There are 13 ranks for each card, and repeating a rank is allowed—even desirable, as it leads to a pocket pair. The number of possible two-card holdings is thus $13 \cdot 13 = 169$.

This quick calculation overcounts the number of possible pairs, for example, a 7/queen pair is counted separately from the same two ranks in the order queen/7. We shall address this overcounting in Section 2.4.3.

As an example of how the FCP is used, we can show that the number of possible events associated with a probability experiment is exponentially greater than the number of elements. Specifically, we have the following result.

Theorem 2.4.2. *If* $\#(\boldsymbol{S}) = n$, *then there are* 2^n *events that may be chosen from* $\boldsymbol{S}$.

Another way to state Theorem 2.4.2 is to say that a set with n elements has 2^n subsets. For example, if we toss a coin, we have $\mathbf{S} = \{H, T\}$. Since we have $\#(\mathbf{S}) = 2$, we expect $2^2 = 4$ events. The complete list of events contained within $\mathbf{S}$ is $\emptyset, \{H\}, \{T\}, \{H, T\}$, which indeed numbers 4.

Proof. Let $A \subset \mathbf{S}$ be an event. For every element $x \in \mathbf{S}$, either $x \in A$ $x \notin A$. That is, there are two possibilities for each element: "Yes, it is in A" or "No, it is not in A." Every arrangement of the n possible yes and no answers corresponds to a different subset of $\mathbf{S}$, and thus to a different event. For example, choosing "no" for every element gives $\emptyset$, and at the other extreme, choosing "yes" each time gives $\mathbf{S}$ itself.

There are n choices to be made, and two options for each of the n choices. Therefore, by the Fundamental Counting Principle, there are

$$\underbrace{2 \cdot 2 \cdot \ldots \cdot 2}_{n \text{ terms}} = 2^n$$

possible events.

Example 2.4.2. If we consider the experiment of drawing a card from a standard deck, then $\#(\mathbf{S}) = 52$ and there are $2^{52} = 4,503,599,627,370$, different events we could consider. Most of these are of limited mathematical interest—for example, the event "Draw any card except the 3♣ or 7♡," with 50 elements, comes up only very rarely.

On the other hand, an event such as "Draw any face card or 10," which has 16 elements, has considerable importance when you've been dealt an ace as your first card in a hand of blackjack.

Example 2.4.3. Old-style mechanical three-reel slot machines often had 20 symbols per reel. When the handle was pulled, each of the reels would spin independently before stopping with one symbol on the central payline. By the Fundamental Counting Principle, there are $N = 20 \cdot 20 \cdot 20 = 8000$ different arrangements of three symbols.

Modern computerized slot machines are considerably more intricate. On a three-reel or five-reel electromechanical slot machine, the position of each reel after the handle is pulled (or the SPIN REELS button is pressed) is determined by a randomly generated number, and the number of symbols on the reels need not bear any direct relationship to the total number of possible random numbers. Some reel configurations, such as a blank on all three reels, may correspond to hundreds of random numbers.

On a video slot machine, the screen displays simulated reels, and the computer can process winning and losing combinations on dozens, even hundreds, of paylines that criss-cross the screen connecting symbols. As with computerized reel slots, a generated random number determines the arrangement of symbols on the virtual reels.

Example 2.4.4. The *Carnival of Mystery* video slot machine, manufactured by International Game Technology, uses a five-reel screen with three symbols displayed on each reel. The game can be played with any symbol on each reel combining with any symbol on each other reel to form a payline. A payline is built by choosing one displayed symbol from the three on each reel, and so the total number of paylines is $3^5 = 243$. On a penny machine, all 243 lines can be played for one wager of 25¢.

A special case of the Fundamental Counting Principle arises when we consider the number of ways to arrange a set of n elements, with no repetition allowed, in different orders. The first element may be chosen in n ways, the second in $n - 1$, and so on, down to the last item, which may be chosen in only one way. The total number of orders for a set of n elements is thus $N = n \cdot (n - 1) \cdot (n - 2) \cdot \ldots \cdot 3 \cdot 2 \cdot 1$. This number is given a special name, *factorial*.

Definition 2.4.1. If $n \in \mathbb{N}$, the *factorial* of n, denoted $n!$, is the product of all of the positive integers up to and including n:

$$n! = 1 \cdot 2 \cdot 3 \cdot \ldots \cdot (n - 1) \cdot n.$$

$0! = 1$, by definition.

It is an immediate consequence of the definition that $n! = n \cdot (n -$ Factorials get very big very fast. $4! = 24$, but then $5! = 120$ and $6! = 720$. $10!$ is greater than 1 million, and $52!$, which is the number of different ways to arrange a standard deck of cards, is approximately 8.066×10^{67}. In the 1970s and 1980s, the largest factorial that could be computed on a standard scientific calculator was $69! \approx 1.711 \times 10^{98}$, a limit well-known to many people whose interest in mathematics was generated in part by playing with early calculators.

Factorials may be computed on the TI-84+ calculator by following these steps:

- Enter n.

- Press $\boxed{\textbf{MATH}}$ to bring up the Math menu.
- Scroll to the right to **PRB**. This brings up the Probability submenu.
- Select option **4: !** and press $\boxed{\textbf{ENTER}}$.
- Press $\boxed{\textbf{ENTER}}$ again.

2.4.2 Where Order Matters: Permutations

Definition 2.4.2. A *permutation* of r items from a set of n items is a selection of r items chosen so that the order matters.

For example, ABC is a different choice of three alphabet letters than CBA. It should be noted that "order" may appear in several forms. One way to determine whether or not order matters in making a selection is to ask if different elements of the selection are being treated differently once they are chosen.

Example 2.4.5. Suppose that a baseball team includes 15 players, each of whom can play any of the 9 positions. If we select nine players to start a game, the assignment of players to positions is a different treatment of the individual players. It would be correct to conclude that the order of selection matters—the same subset of nine players can be assigned to the nine positions in many ways.

We are usually not as interested in a list of all of the permutations of a set as in how many permutations there are. The following theorem allows easy calculation of that number, which is denoted $_nP_r$.

Theorem 2.4.3. *The number of permutations of r items chosen from a set of n items is*

$$_nP_r = \frac{n!}{(n-r)!}.$$

Proof. There are n ways to select the first item. Once an item is chosen, it cannot be chosen again, so the second item may be chosen in $n-1$ ways. There are then $n-2$ items remaining for the third choice, and so on. By the Fundamental Counting Principle, we have

$$_nP_r = n \cdot (n-1) \cdot \ldots \cdot (n-r+1).$$

Multiplying the right-hand expression by $1 = (n-r)!/(n-r)!$ gives

$$_nP_r = n \cdot (n-1) \cdot \ldots \cdot (n-r+1) \cdot \frac{(n-r)!}{(n-r)!}$$
$$= \frac{n \cdot \ldots \cdot (n-r+1) \cdot (n-r) \cdot \ldots \cdot 3 \cdot 2 \cdot 1}{(n-r)!}$$
$$= \frac{n!}{(n-r)!}.$$

The number $_nP_r$ can be calculated easily on a Texas Instruments TI-84+ calculator as follows:

- Enter n.

- Press $\boxed{\textbf{MATH}}$ to bring up the Math menu.

- Scroll to the right to **PRB**. This brings up the Probability submenu.

- Select option **2: nPr** and press $\boxed{\textbf{ENTER}}$.

- Enter r and press $\boxed{\textbf{ENTER}}$.

Example 2.4.6. There are very few casino games in which the order of events matters. One such game, a 1998 promotion at Baldini's Casino in Sparks, Nevada, was described by Barry Meadow in *Blackjack Autumn* [43]. It is common practice for casinos catering to local residents rather than tourists to offer paycheck cashing services, and anyone cashing their paycheck at Baldini's received a free spin on a special video poker machine. This machine offered a million-dollar payout to any player dealt a *sequential* royal flush—the ace, king, queen, jack, and 10 of the same suit, with all five cards in either ascending or descending order.

Receiving a dealt royal flush is difficult enough ($p = \dfrac{4}{2,598,960}$, see Section 2.4.3); to get the cards in order is considerably more difficult. The number of possible arrangements of 5 cards out of 52, where the order matters, is

$$_{52}P_5 = 311,875,200.$$

Of these hundreds of millions of arrangements, only eight are sequential royal flushes (two orders, four suits). The probability of a dealt sequential royal flush, therefore, is

$$\frac{8}{311,875,200} = \frac{1}{38,984,400}.$$

The last denominator is close to the 2013 population of California, making the probability of winning this game about the same as randomly choosing a resident of California and getting the governor.

This was a free promotion, so there was no risk to players, but there was also very little risk to the casino offering this game. (There were other prizes offered; the most common one, according to Meadow, was a free beer.) Of course, if you offer paycheck-cashing services, then you have people in your casino with money in their pockets and a wealth of places where they can spend it, which explains what the casino has to gain from promotions like this one.

Horse Racing

One place where order is important when gambling is in horse racing. The *trifecta* wager requires a gambler to pick the first three horses to finis a race, in order. It's not enough simply to pick the right three horses, it's also necessary to identify which horse wins, which horse comes in second (or *places*), and which is third (the *show* horse). This, of course, requires considerable skill—or luck.

Example 2.4.7. In recent years, the Kentucky Derby, the first race in thoroughbred racing's Triple Crown, has been restricted to 20 horses, determined originally by lifetime winnings and beginning in 2013 by a point system taking the results of preliminary races into account. If you pick your trifecta ticket at random, what is the probability of choosing the winning three horses correctly?

There are

$$_{20}P_3 = 20 \cdot 19 \cdot 18 = 6840$$

different ways to choose the top three horses, so your chance of winning with a single ticket is $\dfrac{1}{6840}$.

Note that the calculation of this probability assumes that all possible topthree outcomes are equally likely, which may not be appropriate to every race. If one horse in the field is very heavily favored (see Example 2.5.3 for one such race), then it may be wise to slot that horse into the winner's spot and consider only the place and show positions for your wager. It may be more sensible, from a purely mathematical perspective, to consider any choice of horses for those two slots as equally likely.

If the trifecta wager seems too risky for your tastes, the *exacta* bet, which requires only that you pick the top two horses in order, may be more suitable. In the example above, with 20 horses starting, the chance of picking the top two finishers in order correctly is $\dfrac{1}{_{20}P_2} = \dfrac{1}{380}$. Going in the other direction, some racetracks offer a *superfecta* wager, which pays off if you correctly pick the top *four* finishers. There are

$$_{20}P_4 = 20 \cdot 19 \cdot 18 \cdot 17 = 116,280$$

different ways to arrange 4 horses out of 20 in order, so your chance of winning on a single superfecta ticket is 1/116,280. It follows that winning superfecta tickets are rather rare and thus are likely to pay off handsomely if the field is relatively large.

How handsomely do winning superfecta tickets pay off? Racetracks use a method called *pari-mutuel* wagering, where the money bet on a given proposition is pooled. As more and more races are run without a winning superfecta wager, the pool accumulates money. The racetrack retains a portion of the wagers, which can run as high as 20%, and the remainder is paid out to the holder of a winning ticket. In managing bets this way, the track assures that its profit is taken off the top before any payoffs are made. Similar separate pools are established for other bets, such as trifecta and exacta bets. The accumulated pool is divided among the winning bettors, so the odds you take when you bet may not be the odds at which you are paid; a winning ticket is paid in accordance with the odds in force at the time that betting for the race ends.

Example 2.4.8. In the 2010 Kentucky Derby, a winning $2 superfecta ticket paid out $202,569.20 [86]. If you wanted to cover every possible permutation of 4 horses from 20, how much money would you have to invest?

At $2 per ticket, your total ticket cost would be $232,560—which is more than the payoff, and so you would win the bet but still lose money.

The challenge to the expert horse bettor is to find information about the horses in the race that suggests that certain outcomes are more or less likely than others—perhaps the favored horse has a history of performing well only in rainy weather, and the forecast for the race is dry. This information can then be used to make bets that are thought to have a better-than-random chance of winning.

Example 2.4.9. If you can safely identify which horse will win, one way to wager is to make a *wheel* bet on the exacta. A wheel combines one horse to win with every one of the other horses to place, and so is guaranteed to win provided that you chose the winner correctly. In a 20-horse race, a wheeled exacta bet is the equivalent of 19 different tickets; a wheeled trifecta bet covers $_{19}P_2 = 342$ different tickets.

2.4.3   Where Order Doesn't Matter: Combinations

Most of the time when gambling, we are not so concerned about the order of events, as when a hand of cards is dealt or a set of lottery numbers is drawn. For counting these arrangements, we are interested in *combinations* rather than permutations.

Definition 2.4.3. A *combination* of r items from a set of n items is a subset of r items chosen without regard to order. The number of such combinations is denoted $_nC_r$.

Here, ABC and CBA are interchangeable combinations, as they are subsets of the alphabet consisting of the same three letters. The different order is not a concern here. If the elements of a selected subset are receiving the same treatment once selected, then the choice is a combination, not a permutation.

Example 2.4.10. The Miss America pageant has 51 entries, one from each state and one from the District of Columbia. If we pick the top four finishers in order, that would be a permutation, since there is a significant difference between finishing first and finishing fourth. If we simply select four contestants to appear at a press conference, then it does not matter which woman is selected first, second, third, or fourth—this is a combination.

Theorem 2.4.4. *The number of combinations of r items chosen from a set of n items is*

$$_nC_r = \frac{n!}{(n-r)! \cdot r!} = \frac{_nP_r}{r!}.$$

Proof. We begin with the formula for the number of permutations:

$$_nP_r = \frac{n!}{(n-r)!}.$$

Since we are looking for combinations, two permutations that differ only in the order of the elements are identical to us. Any combination of r elements from a set of n can be rearranged into $r!$ different orders, by the Fundamental Counting Principle. We then have

$$_nC_r = \frac{_nP_r}{r!}$$
$$= \frac{n!}{(n-r)! \cdot r!},$$

as desired.

This value is often denoted $\binom{n}{r}$, which is read as "n choose r." $_nC_r$ can be calculated on a TI-84+ calculator by following the steps on page 30 for n but replacing **2: nPr** by **3: nCr**.

While exacta, trifecta, and superfecta wagers in horse racing require a gambler to pick the first two, three, or four finishers in order, another wagering option is a box bet. *Boxing* a bet gives the bettor all possible permutations of the selected horses, and thus eliminates the need to pick the exact order correctly. A boxed wager comes with either an increased price or a decreased payoff.

Example 2.4.11. For a 20-horse race, what is the probability of winning a superfecta box wager?

You must select four horses, but the order in which they finish does not matter, which means that your ticket will cover all $_4P_4 = 24$ permutations. Your probability of winning is

$$\frac{_4P_4}{_{20}P_4} = \frac{1}{_{20}C_4} = \frac{1}{4845} \approx 2.06 \times 10^{-4},$$

where the second expression reflects the reality that you are simply trying to pick the right combination of four horses without any regard to order.

Example 2.4.12. The best set of hole cards in Texas Hold'em is a pair of aces, often dubbed "pocket rockets" or "bullets" by players. What is the probability of being dealt a pair of aces?

There are $_4C_2 = 6$ different pairs of aces, and $_{52}C_2 = 1326$ different pairs of cards that can be dealt, since the order does not matter. We find that the probability of pocket aces is

$$p = \frac{6}{1326} = \frac{1}{221}.$$

In the long run, then, you will be dealt a pair of aces once in every 221 hands—which means that if this pair arrives, you should bet it for as long as doing so makes sense, since you have an advantage and this is a genuinely rare event.

Since the formulas for $_nP_r$ and $_nC_r$ involve factorials, which are known to grow very fast, it follows that these numbers also get very big very fast. Starting with a standard 52-card deck, the number of possible subsets of a given size r is quite large, even if order is not considered. Three-card poker (page 111) deals a three-card hand to each player. Since the order does not matter, there are

$$_{52}C_3 = \frac{52!}{49! \cdot 3!} = 22,100$$

different hands. In five-card poker, the number of hands is

$$_{52}C_5 = 2,598,960,$$

and in contract bridge, where each player receives 13 cards, there are

$$_{52}C_{13} = 635,013,559,600$$

hands—over 600 billion.

Example 2.4.13. In five-card draw poker, the royal flush is the highest-ranked hand. The order in which the cards are dealt does not affect the hand's value—indeed, if you are ever dealt a royal flush, it would be bad form to complain that the cards didn't arrive in increasing order. There are 52 cards in a standard deck, and thus there are

$$_{52}C_5 = 2,598,960$$

different 5-card poker hands. Since there are four possible royal flushes—the cards are fixed, so the only variable is the suit—the probability of a royal flush is

$$p = \frac{4}{2,598,960} = \frac{1}{649,740}.$$

In Great Britain, a popular gambling pastime is the football [soccer] pools. Bettors select subsets of 8 games from the list of the weekend's 49 designated games, attempting—if skill plays a part in their choices—to pick the games that will end in draws. Tied games are divided into two subsets: *score draws* where each team scores at least once, and *scoreless draws*, where the final score is 0-0. The leading Pools manager is Littlewoods, which dates back to 1923 and is now part of The Football Pools [27]. Their Classic Pools game invites a player to choose 10–15 games from 49, and covers all possible combinations, or "lines," of 8 games from those selected. A 10-game ticket costs only and thus covers $_{10}C_8 = 45$ possible combinations at a cost of less than 3 pence apiece. Players may select 11 matches for £2.75, covering $_{11}C_8 = 165$ lines, and 12 games, totaling $_{12}C_8 = 495$ lines, for £7.50 [19].

Points are awarded based on the outcomes of the eight games. A game ending in a score draw counts for 3 points, a scoreless draw or void match (one where the match is played on a different date than originally declared, for example) for 2, and a win for 1. Prize funds established from entry fees are divided among bettors whose lines score the top three point totals; if 8 games of the 49 end in score draws, the top prize pool is divided among any players with a line totaling 24 points. The top prize fund typically runs into millions of pounds.

Example 2.4.14. Assuming that exactly eight score draws occur, a player buying a 10-game ticket has probability

$$p = \frac{_{10}C_8}{_{49}C_8} = \frac{45}{450978066} \approx 9.978 \times 10^{-8}$$

of winning a share of the top prize pool.

Many Pools players simply pick their favorite numbers in the range from 1 to 49, thus making this at least as much a game of chance as a game of skill. The Pools are so popular in Britain that in 1963, on a weekend when unusually bad weather led to the cancellation of most games, a "Pools Panel" was appointed to invent outcomes for games so as not to interfere with gambling [27].

The following theorem collects several simple facts about combinations.

Theorem 2.4.5. *For all $n \geq 0$:*

1. $_nC_0 = {_nC_n} = 1$ *and* $_nC_1 = n$.

2. *For all* k, $0 \leq k \leq n$, $_nC_k = {_nC_{n-k}}$.

3. $\sum_{r=0}^{n} {_nC_r} = 2^n$.

Proof.   1. Given a set of n elements, there is only one way to select none of them—that is, there is only one way to do nothing, so $_nC_0 = 1$. Similarly, since the order does not matter, there is only one way to choose all of the items: $_nC_n = 1$.

If we are choosing only one item, we may select any element from among the n, and there are thus n choices possible.

2. We note that every selection of k items from a set of n partitions the set into two disjoint subsets: one of size k and the other of size $n -$ and so choosing k items to take is equivalent to choosing $n - k$ items to leave behind. The conclusion follows immediately.

   Alternately, direct application of the formula for combinations gives the following:

   $$_nC_k = \frac{n!}{(n-k)! \cdot k!} = \frac{n!}{k! \cdot (n-k)!} = \frac{n!}{[n-(n-k)]! \cdot (n-k)!} = {_nC_n}$$

3. If we think of a combination of r items from a set A with $\#(A) = n$ choosing a subset of A with r elements, then the left side of this equation is simply the total number of subsets of A of all sizes.

   From Theorem 2.4.2, we know that the number of subsets of a set with n elements is 2^n. Since we have counted the set of all subsets of A two different ways, those two expressions must be the same, completing the proof.

Poker

Example 2.4.15. In computing the probability of various poker hands, we need to account for every card in the hand—for example, when computing the probability of two pairs, we need to exclude the possibilities that the two pairs are of the same rank, which would mean that the hand was four of a kind, and that the fifth card is the same rank as one of the pairs, which would turn the hand into a full house.

The form of a two-pair hand is $AABBX$, where A, B, and X are card ranks, $A \neq B$, and neither A nor B is equal to X. A and B, together, may be chosen in $_{13}C_2$ ways, since the order does not matter. Once the ranks have been chosen, the pairs can consist of any two of the four cards of those ranks, and the order in which we pick them doesn't matter. By the Fundamental Counting Principle, we can find the two pairs in $_{13}C_2 \cdot (_4C_2)^2 = 78 \cdot 36 = 2808$ ways.

The fifth card, X, can be any of the 48 remaining that do not match either pair. For example, if we have a pair of queens and a pair of 7s, the fifth card can be any card remaining except for the two queens and the two 7s, and there are 44 of these.

Putting this all together with the Fundamental Counting Principle and remembering that the number of possible poker hands is $_{52}C_5 = 2,598,960$, we have

$$P(2 \text{ pairs}) = \frac{2808 \cdot 44}{2,598,960} = \frac{123,552}{2,598,960} \approx .0475 = 4.75\%.$$

Example 2.4.16. In the hierarchy of poker hands, a flush beats a straight. This is an ordering that people often get backwards, because to the casual observer, getting five cards in sequence but not of the same suit seems very similar to getting five cards of the same suit but not in sequence. By counting the number of possible hands of each type, the reason for this ordering becomes clear.

Flushes are easier to count. We select a suit, in $4 = {}_4C_1$ ways, and then choose 5 of the 13 cards in that suit. Multiplication gives us

$$4 \cdot ({}_{13}C_5) = 5148$$

flushes.

This number includes straight flushes and royal flushes, which a poker player would not count as a flush. To obtain an accurate count, we must subtract these hands from our total. Any card from ace through 9 may be the lowest-ranking card in a straight flush, and so there are 36 of those. For royal flushes, the only variable is the suit, so there four royal flushes. Subtracting these 40 hands from 5148 leaves 5108 flushes.

To count straights, while the order in which the cards are dealt doesn't matter, we can facilitate the count by thinking of the cards in numerical order. There are 40 choices for the lowest card in a straight, because an ace can be either high or low, so all cards from ace through ten are candidates. Having chosen this first card, the remaining ranks are determined, and we have four choices for each of the four remaining cards, giving

$$40 \cdot 4^4 = 10,240$$

possible straights. Once again, we need to subtract the 40 straight and royal flushes, leaving 10,200 hands that are called a straight.

Comparing these numbers, we see that there are almost exactly twice as many possible straights as flushes, which accounts for the relative ranking of the two hands.

Example 2.4.17. Most Texas Hold'em experts advise a fairly conservative betting strategy, folding most initial hands unless they contain a pair, two high cards (such as AK), or certain suited *connectors*, which are two hole cards of consecutive ranks, so called because of their potential to yield a straight. According to this reasoning, which is backed up by millions of simulated hands, a pair of hole cards such as 9♠ 2♠ should be folded immediately.

Some novice players note "But they're *suited!*" and insist on betting that sort of hand. What is the probability of drawing into a flush starting with these two cards?

To get a flush in this example, the flop, turn, and river must contain at least three spades among their five cards. In counting the number of spades in a set of five cards, we need to account for the suits of all five cards, so that a set with four spades is counted with four-spade sets and not also with the three-spade sets. For example, if we wish to count the number of ways to draw exactly three spades, we must count the number of ways to draw three spades, count the number of ways to draw two nonspades, and then use the Fundamental Counting Principle to multiply those two numbers. Given our hole cards, there are 11 spades and 39 other cards remaining to be dealt; it follows that the number of ways to draw five cards containing exactly three spades is

$$_{11}C_3 \cdot \ _{39}C_2.$$

Accounting for every card, we have

$$\underbrace{_{11}C_3 \cdot \ _{39}C_2}_{\text{3 spades}} + \underbrace{_{11}C_4 \cdot \ _{39}C_1}_{\text{4 spades}} + \underbrace{_{11}C_5}_{\text{5 spades}} = 165 \cdot 741 + 330 \cdot 39 + 462 = 135,597$$

ways for the community cards to complete a flush.

The probability of drawing into a flush, therefore, is

$$p = \frac{135,597}{_{50}C_5} = \frac{135,597}{2,118,760} \approx .0640 \approx \frac{1}{16},$$

only about a 6.4% chance of success including the very low probability of drawing into a straight flush. Betting on this sort of hand in the hope of drawing a flush is a losing proposition.

There is an exception to this betting vs. folding strategy. Each round of Texas Hold'em begins with two designated players making *ante*, or starting, bets before the cards are dealt. These bets, whose amounts are fixed, are called the *small blind* and the *big blind*, which is typically twice the small blind. These bets are often thought of as an initial bet and an initial raise. The players charged with making these bets are identified by a button that rotates clockwise around the table [62].

If it is your turn to make the big blind bet, and no further bets are made in the opening round before the flop, there is no cost to you to remain in the game and see the flop, and thus there is no reason why you should fold any hand, no matter how weak. While the chance of improving two weak hole cards into a winning hand is slight, there is nothing lost in this case in seeing what cards come out on the flop.

Lotteries

An *r/s lottery* is a game where players bet on which r numbers will be drawn from a set of s, and games operating on this basic premise are common throughout the USA. Lotteries are another example of a game of chance where the order of events, in this case the drawing of numbers, does not matter. We are perhaps most familiar with multimillion-dollar lottery prizes that attract media attention, but the typical r/s lottery offers prizes other than the life-changing jackpot.

Example 2.4.18. The Michigan State Lottery's "Classic 47" game is a 6/47 lottery, with the following prize structure:

Match	Payoff
3	$5
4	$100
5	$2500
6	Jackpot

The jackpot starts at $1 million and increases each week that it is not won, resetting to $1 million in the drawing after a winner is determined. What ar the probabilities of winning each of the four prizes?

The probability of winning each of these amounts is computed using the formula for combinations, recognizing that the order in which the numbers are drawn is not important, only that the numbers you've chosen be drawn somewhere in the set of six. In reporting the winning numbers for a lottery drawing, media outlets will arrange them in ascending order, but this is a convenience and almost certainly not a faithful representation of the order of events. When doing so, it is important to account for each of the six numbers being drawn in both the numerator and denominator.

The probability of matching exactly k numbers out of 6, chosen from a set of 47, is given by the formula

$$P(\text{Match } k) = \frac{{}_6C_k \cdot {}_{41}C_{6-k}}{{}_{47}C_6}.$$

In the numerator, the first term counts the number of ways to choose the winning numbers, while the second is the number of ways to choose the nonwinning numbers from among the 41 not appearing on the ticket. When computing these numbers on a calculator, you should use the built-in n routine and not round off any results until the end of the calculation. For $k = 3, 4, 5$, and 6, we have the following probabilities:

k	$P(\text{Match } k)$
3	.0199
4	.0011
5	.00002291
6	.00000009313

More generally, the probability of matching k of the numbers drawn in an r/s lottery is

$$P(\text{Match } k) = \frac{{}_rC_k \cdot {}_{s-r}C_{r-k}}{{}_sC_r}.$$

Example 2.4.19. Powerball is a popular game administered by the Multi-State Lottery Association. Tickets for this lottery are sold in 43 states, the District of Columbia, and the U.S. Virgin Islands. The game involves choosing five white balls from a bin containing balls numbered from 1 to 59 and one red ball from a separate bin with balls numbered from 1 to 35—this last ball is known as the Powerball. The jackpot is won by a player holding a ticket whose numbers match those chosen by the lottery operators. The record Powerball jackpot of $590.5 million was awarded in May 2013.

Order doesn't matter when choosing the white balls, so they may be chosen in $_{59}C_5 = 5,006,386$ different ways. By the Fundamental Counting Principle, we must multiply this number by the 35 choices for the Powerball; this gives 175,223,510 possible results for each drawing, and so the probability of winning the jackpot on a single ticket is $\dfrac{1}{175,223,510}$.

To put this number in perspective, consider Michigan Stadium at the University of Michigan-Ann Arbor, which is the largest stadium in the USA, with a capacity of 109,901. If we were to fill the stadium 1594 times, we would have almost as many people as there are possible Powerball tickets. In the history of the stadium, which includes many years before its capacity was expanded to its current number, there have not been even half that many games. In short, fewer people have attended Michigan home football games than there are Powerball combinations.

Keno

The same mathematical principles underlying lotteries can be used in a casino to analyze *keno*. Keno is one of the easiest casino games to understand and to play. In a simple keno game, the player chooses anywhere from 1 to 20 numbers in the range from 1 to 80. The casino then selects 20 numbers from this set. Originally this was done with a cage or blower containing numbered balls—called a *goose*—but many casinos now use a computer to generate the numbers. The player wins if a certain number of the numbers he or she chose appears among the casino's selection. In many casino resorts, the keno games run continuously, and the numbers can be found on video monitors throughout the premises or on the hotel's closed-circuit television channels, so players can monitor their wagers from the restaurants or their hotel rooms. Some casinos broadcast their drawings over the Internet (one such Web site is kenousa.com) so the games can be tracked by players even when they aren't in the casino. Since some keno operations allow bets on 1000 consecutive games that can take over three days to complete, this is convenient for people who place such a wager.

There are $_{80}C_{20} = 3,535,316,142,212,174,320$ ways for the casino to draw its 20 numbers. This will be the denominator in our keno calculations. In the history of keno, there is no record of any player who has matched all 20 numbers drawn by the casino on a single ticket, an occurrence which has probability 1/3,535,316,142,212,174,320, approximately 2.829×10^{-19}. It is more likely that you would toss a coin 61 times and get tails on every toss, an event with probability $p = 2^{-61} \approx 4.337 \times 10^{-19}$. If we imagine one keno drawing every second since the Big Bang, approximately 15 billion years ago, this would only account for 4.7335×10^{17} combinations, and the probability of any single combination being drawn in that time is only about 13.4%.

Possibly in recognition of this mathematical fact, the Colorado Belle Casino in Laughlin, Nevada offers a 20-number keno game which, for a bet, pays off the same $100,000 if the ticket matches 15 or more numbers. Since the probability of hitting 15 numbers out of 20 is small (see Exercise 2.14), and the chances of hitting 16 through 20 are smaller still, this is no great advantage to the player.

Many casinos offer a wide range of keno bets. For example, the players' guide for keno at the Soaring Eagle Casino in Mount Pleasant, Michigan includes 23 different games. The following pay table is for a $1 bet on their "Mark 7 Numbers" game.

Outcome	Payoff
Match 4	$1
Match 5	$20
Match 6	$300
Match 7	$8000

If a player bets more than $1 (the maximum bet is $5), these payoffs are increased by multiplying each payoff by the amount wagered.

Example 2.4.20. What is the probability of matching four out of seven numbers and qualifying for the lowest prize (which is a simple refund of the ticket price)?

There are $_7C_4 = 35$ ways to select the four numbers that are matched by the casino. We must also account for the other 16 numbers, and they must be chosen from the 73 numbers that the player did not select. This can be done in $_{73}C_{16} = 5,271,759,063,474,612$ ways. By the fundamental counting principle, any subset of 4 matches can be combined with any subset of 16 non-matches to form a winning combination, and so we multiply these two numbers together to find the numerator: there are 184,511,567,221,611,420 (over 184 quadrillion) ways to win this bet. That sounds encouraging until we compare it to the denominator, which is 3,535,316,142,212,174,320—over 3.5 quintillion, and nearly 20 times larger.

$$P(\text{Match } 4) = \frac{_7C_4 \cdot {}_{73}C_{16}}{_{80}C_{20}}$$
$$= \frac{184,511,567,221,611,420}{3,535,316,142,212,174,320} \approx .0522.$$

The probabilities of the other winning events in the table can be calculated similarly, using the general formula

$$P(\text{Match } k) = \frac{{}_7C_k \cdot {}_{73}C_{20-k}}{{}_{80}C_{20}}.$$

For the remaining money-winning combinations, this formula gives the following probabilities:

$$P(\text{Match } 5) = \frac{{}_7C_5 \cdot {}_{73}C_{15}}{{}_{80}C_{20}} \approx .0086.$$

$$P(\text{Match } 6) = \frac{{}_7C_6 \cdot {}_{73}C_{14}}{{}_{80}C_{20}} \approx .00073.$$

$$P(\text{Match } 7) = \frac{{}_7C_7 \cdot {}_{73}C_{13}}{{}_{80}C_{20}} \approx .00002440.$$

Example 2.4.21. Making only a single n-number bet is not terribly interesting for a gambler, nor is it lucrative for the casino. By making a *way* bet, it is possible to cover more combinations of numbers, frequently at a reduced price per wager, and increase the probability of drawing into a winning combination. To make a way bet, the bettor selects several groups of numbers, as in Figure 2.3, where three groups of four numbers, indicated by shaded blocks, have been chosen.

1	2	3	4	5	6	7	8	9	10
11	12	13	14	15	16	17	18	19	20
21	22	23	24	25	26	27	28	29	30
31	32	33	34	35	36	37	38	39	40
41	42	43	44	45	46	47	48	49	50
51	52	53	54	55	56	57	58	59	60
61	62	63	64	65	66	67	68	69	70
71	72	73	74	75	76	77	78	79	80

FIGURE 2.3: Keno bet slip with shaded 3/4, 3/8, 1/12 way bet

It should be noted that the numbers comprising a block need not be adjacent as shown here; by placing an X through the desired numbers and circling the entire set, any group of numbers may be combined into a block. The player may then specify how the blocks are to be combined. For the ticket shown in Figure 2.3, there are three 4-spot bets consisting of the three individual blocks, three 8-spot tickets, which arise from combining every pair of 4-spot blocks into a "virtual" 8-spot block, and a single 12-spot ticket encompassing all of the selected numbers. If all seven ways were then played, this ticket would be called a 3/4, 3/8, 1/12 seven-way keno ticket.

At the Soaring Eagle, each way may be purchased for as little as 10 provided that the ticket as a whole contains at least \$3 of bets. A way ticket provides for the possibility of multiple wins, but this possibility is purchased with a greater cost for the ticket as a whole.

Example 2.4.22. The blocks on a way bet need not be the same size. Consider the ticket shown in Figure 2.4, which has seven blocks: three 2-spots and four 4-spots.

1	2	3	4	5	6	7	8	9	10
11	12	13	14	15	16	17	18	19	20
21	22	23	24	25	26	27	28	29	30
31	32	33	34	35	36	37	38	39	40
41	42	43	44	45	46	47	48	49	50
51	52	53	54	55	56	57	58	59	60
61	62	63	64	65	66	67	68	69	70
71	72	73	74	75	76	77	78	79	80

FIGURE 2.4: Keno bet slip with shaded 3/2, 7/4, 13/6, 18/8, 22/10 way bet

Suppose now that the 20 numbers drawn against this ticket are the following, with the covered numbers bolded:

$$6, \mathbf{7}, 15, 19, 21, 23, \mathbf{24}, 26, 28, 39,$$
$$\mathbf{44}, 45, 48, 49, \mathbf{51}, 64, \mathbf{67}, 69, 71, 73$$

How many wagers are formed by the selected blocks, and what is the net payoff on this way ticket?

These blocks may be combined in the following ways to form virtual tickets with ten or fewer spots:

- The four 2-spots may be played individually.

- There are seven 4-spot tickets: the four 4-spots played individually and $_3C_2 = 3$ arising from all possible combinations of the three 2-spot blocks: {37,38,51,61}, {37,38,67,77}, and {51,61,67,77}.

- Each 4-spot block may be combined with each of the 2-spot blocks to form a separate 6-spot block, and the 3 2-spot blocks may be combined for a virtual 6-spot ticket, making 13 6-spot choices.

- There are 18 8-spot blocks:

    - $_4C_2 = 6$ from combining two 4-spot blocks.
    - $_3C_2 \cdot 4 = 12$ from combining any two 2-spot blocks with a 4-spot block.

- Finally, there are 22 10-spot blocks:

- Choose any two 4-spot blocks ($_4C_2 = 6$ ways) and add in any of the three 2-spot blocks, for a total of 18.

- Use all three of the 2-spot blocks with any one 4-spot block, totaling four more.

This results in 63 different combinations of the seven blocks. Suppose that you wager the ticket in Figure 2.4 at this 10¢ minimum, which represents a total investment of $6.30. The payoff table, which is shown in Table 2.5, is taken from the casino's $1 wager paytable.

TABLE 2.5: Paytable for a $1 keno wager, Soaring Eagle Casino

Match	Number of spots marked				
	2	4	6	8	10
2	$12	$1			
3		$3	$1		
4		$125	$4		
5			$90	$10	$2
6			$1499	$90	$20
7				$1499	$130
8				$15,000	$1000
9					$8000
10					$25,000

The $1499 payoff amounts, which may seem strange, have been specifically chosen because a keno payoff of $1500 or more must be reported by the casino to the Internal Revenue Service. (All gambling winnings are judged by the IRS to be taxable income, whether reported directly by the casino or not. Gambling losses, if carefully documented, may be declared as a deduction, but only up to the extent of your winnings.) It is easy to see that multiple wins at the lower levels, where they are most likely, may be necessary to win back the $6.30 cost of the original ticket.

The only winning combinations are from the 4-spot 14, 24, 34, 44, and the combined 2-spots 51, 61, 67, 77, which each return $1 of the wager. The net return on this ticket in this drawing is –$4.30.

Example 2.4.23. Perhaps the ultimate way bet ticket is a *king* ticket. A single number played as part of a way bet is called a *king*, and a keno ticket consisting entirely of kings provides a multitude of betting combinations. A ticket with seven kings circled and all combinations activated has 127 possible winning ways. This number is $2^7 - 1$, and may be easily calculated without resorting to a list by referring to Theorem 2.4.2: There are 2^7 ways to choose a subset of the seven kings, but since one of these, the empty set, does not correspond to a keno combination, we subtract 1 from this number.

If all 7 numbers are drawn, an event with probability

$$p = \frac{{}_{73}C_{13}}{{}_{80}C_{20}} = \frac{86,270,750,308,176}{3,535,316,142,212,174,320} \approx 2.440 \times 10^{-5},$$

the \$12.70 ticket will pay off 127 times, for a total of \$35,811.

Example 2.4.24. The Soaring Eagle Casino also offers a gimmick keno bet called the "Outside Eagle's Edge" bet. This bet allows you to bet on the 32 edge numbers on the betting slip, which are shaded in Figure 2.5.

1	2	3	4	5	6	7	8	9	10
11	12	13	14	15	16	17	18	19	20
21	22	23	24	25	26	27	28	29	30
31	32	33	34	35	36	37	38	39	40
41	42	43	44	45	46	47	48	49	50
51	52	53	54	55	56	57	58	59	60
61	62	63	64	65	66	67	68	69	70
71	72	73	74	75	76	77	78	79	80

FIGURE 2.5: Keno bet slip with Outside Eagle's Edge bet shaded

Among the 16 payoffs, you win \$200 (reduced from \$25,000 in September 2011) if none of the edge numbers is selected. What is the probability of this happening?

The calculation is actually quite simple: You win if the 20 numbers chosen by the casino are among the 48 that are *not* on the edge of the bet slip. The probability of this is

$$P = \frac{{}_{48}C_{20}}{{}_{80}C_{20}} = \frac{16,735,679,449,896}{3,535,316,142,212,174,320} \approx 4.73 \times 10^{-6} \approx \frac{1}{211,244}.$$

In reducing the payoff on this event from \$25,000 to \$200, the casino was almost surely reacting to a nonexistent threat. Provided that the first winning ticket doesn't happen too soon after the game is launched, a casino can easily afford to offer this bet—if only one bet in 211,244 wins, and if each bet is for \$1, then in the long run, the casino has made a profit of \$186,244 after paying a \$25,000 winner.

2.5 ODDS

In the world of gambling, probabilities are often encountered in terms of *odds*.

Definition 2.5.1. The *odds against* an event A is the ratio $P(A') : P(A)$, or $P(A')/P(A)$.

This is often stated in a form like "x to 1," as is the case in horse racing, for example. Most of the time when odds are quoted, they are odds against. We can also consider the odds in favor of an event.

Definition 2.5.2. The *odds for* an event A is the ratio $P(A) : P(A')$—the reciprocal of the odds against A.

Example 2.5.1. When rolling 2d6, *snake eyes* is the name given to a roll of 1-1. Since there are 36 ways for the dice to land, and only one is a 1-1, the probability of snake eyes is 1/36. The odds against snake eyes would then be

$$\frac{35}{36} : \frac{1}{36},$$

or 35 to 1.

Example 2.5.2. The odds against the double street bet (six numbers) at American roulette are

$$\frac{32}{38} : \frac{6}{38},$$

or 32 to 6, which reduces to $5\frac{1}{3}$ to 1.

Note that the payoff for the double street bet is 5 to 1, so this bet is paid off at less than its true odds of $5\frac{1}{3}$ to 1, which is where the casino gets its advantage. Contrary to common belief, a casino doesn't make its money from wagers collected from losing bettors, but rather from paying off winners at less than true odds.

Example 2.5.3. At the 1973 Belmont Stakes, the final race in thoroughbred racing's Triple Crown, the quoted odds on the favored horse, Secretariat, were 1 to 10. Using this information, we can compute $P(A)$, the probability that the oddsmakers assigned to Secretariat winning:

$$\frac{1}{10} = \frac{P(A')}{P(A)} = \frac{1 - P(A)}{P(A)}.$$

Cross-multiplying gives

$$P(A) = 10 - 10 \cdot P(A)$$
$$11 \cdot P(A) = 10$$
$$P(A) = \frac{10}{11} \approx .909 = 90.9\%.$$

Secretariat won the race by 31 lengths, becoming the first Triple Crown winner since Count Fleet in 1948. Both the margin of victory and the time, 2:24.00, remain Belmont Stakes records.

We can rearrange the formula for odds and derive the following result.

Theorem 2.5.1. *If the odds against an event A are x to 1, then P(A*
$$\frac{1}{x+1}.$$

Proof. We have

$$\frac{1 - P(A)}{P(A)} = \frac{x}{1}.$$

Cross-multiplying gives

$$1 - P(A) = x \cdot P(A),$$

or

$$(x+1) \cdot P(A) = 1.$$

Dividing by $x + 1$ gives

$$P(A) = \frac{1}{x+1},$$

completing the proof.

Applying this result in a casino sports book can lead to some interesting revelations.

Example 2.5.4. In the summer of 2012, the New York-New York Casino in Las Vegas set future odds on each National Football League team's chance of winning Super Bowl XLVII in February 2013. Odds ranged from 9 to 2 $\left(P(\text{Win}) = \frac{2}{11}\right)$ on the San Francisco 49ers up to 150 to 1 $\left(P(\text{Win}) = \frac{1}{151}\right)$ the Jacksonville Jaguars. If we were to place a $10 bet on each of the 32 teams, we would invest $320 and only make a profit if the odds on the champion at the time of our wager were at least 32 to 1 (our $10 wager on the winning team would be returned with our payoff). Only 14 of the 32 teams had odds running at least this high, and those, of course, were judged to be the weakest teams in the league. The casino's cumulative probability of 1 of those 14 teams winning the Super Bowl was, at that time, a mere .2269.

According to Axiom 2, we would expect the sum of the 32 associa probabilities to be 1; it was

$$\frac{49,243,234,974,591,245}{26,687,953,298,243,952},$$

approximately 1.8451.

How are we to interpret this deviation from the laws of probability? The extra .8451 reflects the amount of profit that the casino expects to make on

Super Bowl futures bets. New-York-New York has declared its best estimate of the probability of each team winning the Super Bowl in setting its odds. If gamblers bet on each NFL team in proportion to the odds assigned by New York-New York, then $\frac{2}{11}$ of the wagered money ($18.18 out of each $100) will be bet on San Francisco and $\frac{1}{151}$ ($.66 of every $100) will be bet on Jacksonville, for example.

The *payoff* on one of these bets will be approximately $100, regardless of which team wins. (Payoffs are rounded down to the nearest cent or 10 always in the casino's favor.) Since the 32 probabilities add up to more than 100%, the casino will take in more than $100 for every $100 it eventually pays out to the winners. Specifically, the casino expects to collect $184.51 for every $100 in winnings it will pay.

This excess is the fraction

$$\frac{.8451}{1.8451} \approx .4580,$$

which denotes the "over-round," or proportion of the total money wagered that will *not* be returned to winning gamblers, but will be retained by the casino as profit [13]. Here, the casino expects to hold about 45.8% of the total amount bet.

New York-New York is on sound financial footing in offering these wagers. Super Bowl XLVII was won by the Baltimore Ravens, on whom the casino had placed 18 to 1 odds.

2.6   EXERCISES

Answers begin on page 253.

2.1. Some casinos pay a small bonus on certain blackjack hands—for example, a hand containing three 7s, totaling 21. In a double-deck blackjack game, find the probability of being dealt three 7s.

2.2. Find the probability of being dealt a pair as your hole cards in Texas Hold'em.

2.3. Remembering that an ace can count as either high or low, find the probability that your hole cards form a pair of connectors.

2.4. Some local variations of poker recognize a hand called a *blaze*, which consists of five face cards. (Note that only jacks, queens, and kings are considered face cards.) A blaze beats two pairs but loses to three of a kind, although a blaze containing three of a kind or higher (for example, four queens and a jack) need not be called a blaze [20, p. 25].

a. Find the probability of being dealt five face cards.

b. Find the probability that a player is dealt a hand that he or she calls a blaze—which must consist of two pairs of face cards and a fifth face card of the third rank.

2.5. Find the probability that a five-card poker hand contains five cards of the same color.

2.6. Find the probability that a poker hand is a full house: three of a kind plus a pair.

2.7. Calculate the probability of the poker hand three of a kind. Note that in a hand holding three of a kind, if the other two cards form a pair, the hand is a full house; these need to be excluded from your calculations.

2.8. Consider a modified deck of cards containing six suits, with the usual 13 cards in each suit for a total of 78 cards. The two new suits are red *crowns* and black *anchors*.

a. In a five-card poker game played with this deck, five of a kind would be a possible hand. Compute the probability of five of a kind. Would five of a kind outrank a royal flush?

b. Find the probability of being dealt a hand containing five cards of different suits, which is called a *rainbow*.

c. A *rainbow straight* is a hand consisting of five cards in sequence and all of different suits. Find the probability of a rainbow straight.

2.9. The Outside Eagle's Edge keno bet (Example 2.4.24) offers a payout of $100,000 if 20 of the 32 edge numbers are chosen by the casino. Find the probability of this payoff.

2.10. Calculate the number of betting combinations on a keno way ticket consisting of two 2-spots, four 3-spots, and a 5-spot, with the understanding that no combination may involve more than 15 numbers.

2.11. The Kewadin Casinos in Sault Sainte Marie and St. Ignace, Michigan offer a way bet called the "Cover All" keno wager. As the name suggests, the Cover All bet has all 80 numbers working for the player. In one Cover All game, the numbers are divided into 2×2 blocks of 4, according to how they are arranged on the bet slip (see Figure 2.3), so 1-2-11-12 is one block, as is 3-4-13-14, and so on up to 69-70-79-80. These 20 blocks are each combined with every other block to form virtual 8-spot tickets. Show that there are 190 distinct 8-spot combinations available in a Cover All game.

2.12. Suppose that on a football (soccer) weekend in Great Britain, 11 of the 49 games end in score draws. What is the probability that your ticket for ten games includes at least one line with eight score draws?

2.13. The "Top and Bottom" keno game at the Kewadin Casinos does not require the player to choose any numbers. By playing a $5 Top and Bottom ticket, you win if 13 or more of the 20 numbers drawn fall on the top half of the ticket, which bears numbers from 1 to 40, or on the bottom half, where the numbers run from 41 to 80. Prizes are awarded according to how many numbers hit: a 13-7 or 7-13 top/bottom split pays off with a free ticket for another game, while a 20-0 or 0-20 split pays $25,000.

Find the probability of winning on a Top and Bottom ticket.

2.14. Recall the Colorado Belle's $20 keno wager, whose top prize of $100,000 is won when the player matches 15 or more of the 20 drawn numbers.

a. Find the probability of hitting exactly 15 of 20 numbers in keno.

b. By mimicking the calculation in part a, find the probability of matching 16, 17, 18, 19, and 20 numbers.

2.15. *2by2* is a game offered by the Multi-State Lottery Association, the organization responsible for Powerball, in Kansas, Nebraska, and North Dakota. 2by2 is effectively a double 2/26 lottery: players pick two numbers, in the range 1 to 26, in each of two colors: red and white. Two numbers in each color are drawn, and a ticket wins if at least one of its four numbers matches them. The payoff table for 2by2 is shown in Table 2.6.

TABLE 2.6: 2by2 payoff table [1]

Red matches	White matches	Prize
2	2	$22,000
2	1	$100
1	2	$100
2	0	$3
0	2	$3
1	1	$3
1	0	Free ticket
0	1	Free ticket

Due to symmetry, there are five different winning probabilities. Find each one.

Compound Events

3.1 ADDITION RULES

Our next challenge will be to extend our understanding of probability to *compound* events: events that can be broken down into several simple events. We can find the probability of these simple events using techniques from Chapter 2; this chapter allows us to combine those probabilities correctly to find probabilities of more complicated events.

Definition 3.1.1. Two events A and B are *mutually exclusive* if they have no elements in common—that is, if they cannot occur together.

Example 3.1.1. If we draw one card from a deck and record its suit, the events $A = \{\text{The card is a diamond}\}$ and $B = \{\text{The card is a spade}\}$ mutually exclusive.

Example 3.1.2. Again drawing one card from a deck, the events $A = \{\text{The card is a diamond}\}$ and $C = \{\text{The card is a 7}\}$ are not mutually exclusive as there is one card, the $7\Diamond$, common to both events.

In computing probabilities, we may be in a situation where we know P and $P(B)$ and want to know the probability that either A or B occurs. In terms of set theory, this is asking for $P(A \cup B)$. The addition rules described next allow us to compute this new probability in terms of the known ones.

Theorem 3.1.1. *(The First Addition Rule)* If A and B are mutually exclusive events, then

$$P(A \cup B) = P(A) + P(B).$$

The First Addition Rule is, of course, just Axiom 3 with $n = 2$. If A and are not mutually exclusive, a slightly more complicated formula can be used to calculate $P(A \cup B)$.

Theorem 3.1.2. *(The Second Addition Rule)* If A and B are any two events, then

$$P(A \cup B) = P(A) + P(B) - P(A \cap B).$$

Proof. By definition,

$$P(A \cup B) = \frac{\#(A \cup B)}{\#(\mathbf{S})}.$$

What we need to do is compute $\#(A \cup B)$. Elements of $A \cup B$ can be counted by adding together the number of elements of A and of B, but if they belong to both, they have just been counted twice. In order that each element is only counted once, we must subtract out the number of elements that belong to both A and B, that is, the number of elements in $A \cap B$. This gives

$$\#(A \cup B) = \#(A) + \#(B) - \#(A \cap B).$$

Dividing by $\#(\mathbf{S})$ completes the proof.

We can see that the First Addition Rule is a special case of the second, for if A and B are mutually exclusive, then $A \cap B = \emptyset$ and thus (by Theorem 2.2.1) $P(A \cap B) = 0$.

Example 3.1.3. Suppose that you are playing five-card draw poker and have been dealt the following hand:

$$4 \spadesuit \; 5 \spadesuit \; 6 \spadesuit \; 7 \spadesuit \; K \heartsuit.$$

If you discard the king, what is the probability of your drawing a card that will complete either the straight or the flush?

The question asks for your chance of drawing either a spade, a 3, or an 8. Define the following events:

$$F = \{\text{Draw a spade}\}$$
$$S = \{\text{Draw a 3 or an 8}\}$$

The event F completes a flush, while the event S completes a straight. The intersection of these two events is $F \cap S = \{3\spadesuit, 8\spadesuit\}$, and consists of the cards that will complete a straight flush.

There are 47 unknown cards remaining in the deck. Of these, nine are spades and eight are either 3s or 8s. It follows that $P(F) = \frac{9}{47}, P(S) = $ and $P(F \cap S) = \frac{2}{47}$, and thus

$$P(F \cap S) = \frac{9}{47} + \frac{8}{47} - \frac{2}{47} = \frac{15}{47} \approx .3191,$$

so you have slightly more than a 30% chance of drawing to complete a straight or flush.

3.2   MULTIPLICATION RULES AND CONDITIONAL PROBABILITY

Definition 3.2.1. Two events A and B are *independent* if the occurrence of one has no effect on the occurrence of the other one.

Two events that are mutually exclusive (Section 3.1) are explicitly independent, since the occurrence of one eliminates the chance of the other occurring. Moreover, two events that are independent cannot be mutually exclusive.

Example 3.2.1. Consider a simple experiment where you toss a fair coin and simultaneously roll a fair d6. Unless you're deliberately trying to bounce one object off the other, the two throws can reasonably be said not to have any influence on each other. We conclude that the results of the tosses are independent events.

Example 3.2.2. It is a fundamental principle of gambling mathematics that *successive trials of random experiments are independent*. This includes successive die rolls at craps, successive wheel spins at roulette, and successive weekly drawings of six lottery numbers, but *not* successive hands in blackjack—for in blackjack, a card played in one hand is a card that cannot be played in the next hand. Since the composition of the deck has changed, we are not considering successive trials of the same random experiment.

This principle is not always well-understood by gamblers, and the inability or unwillingness to understand the doctrine of independent trials is sometimes called the *Gambler's Fallacy*. This fallacy is commonly committed by roulette players who have too strong a belief in the Law of Large Numbers, although it can crop up in any casino game where successive trials are independent.

Example 3.2.3. A frequently observed form of the Gambler's Fallacy comes when roulette players rush to bet on red after a long string—three or four—of successive black numbers has come up, on the grounds that red is "due" so that things will "even up."

Dice, cards, and roulette wheels don't understand the laws of probability. They have no knowledge of the mathematics we humans have devised to describe their actions, and they certainly don't understand what the long-term distribution of results is supposed to be. For the same reason, in Example 3.2.3, it would be equally erroneous to flock to black on the grounds that black is "hot."

Casinos are willing to cater to these mistaken beliefs. At many casinos, the roulette tables are equipped with lighted signs that automatically detect and display the results of the last 20 or so spins. Although the previous spins of the wheel have no effect on the next spin (unless the wheel is defective), if the players wish to track them, identify spurious patterns, and bet accordingly, the casino will gladly accommodate them by providing these results. These signs cost around $7000; one casino installing them found that they paid for themselves in the form of increased roulette income within six weeks [58].

Example 3.2.4. Lucrative though these signs may be for the casino, there is no guarantee that the numbers reported on them are accurate, and a careful reading of the fine print on a casino's roulette information brochure will reveal that no guarantee of accuracy is made. This was pointed out in dramatic fashion in June 2012, when an American roulette board at the Rio Casino in Las Vegas showed seven straight 19s. Since the individual spins are independent, the probability of getting the same number seven times in a row is

$$p = \left(\frac{1}{38}\right)^6 = \frac{1}{3,010,936,384}.$$

In computing this probability, the first spin can be any number; we are merely computing the probability that the next six spins match it; hence the exponent in the equation above is 6, not 7. While it is within the realm of reason that some roulette wheel somewhere in the world will, given an amount of time less than the average human lifetime, come up on the same number seven times in a row, it was revealed only a couple of days later that the display—though not the wheel itself—had been malfunctioning.

Recall that the payoff on a straight number bet is 35 to 1. Assuming that the display was accurate, had you bet $1 on 19 before the first spin and then let your winnings ride through the next seven spins—and had the foresight to withdraw your windfall *before* the next spin—you would have won $35 on the first spin, then $36 · 35 on the second spin, and in general, $35 · 36^{n-1}$ on spin #n. After seven spins, your total winnings would amount to $78,364,164,095: over 78 *billion* dollars. On a practical level, the house limits on maximum be size would have prevented you from letting your winnings ride more than two or three times. For example, if the casino's limit on "inside bets," such as bets on a single number, is $100 (see Figure 7.1 on page 223), you would be unable even to place the third bet in that sequence, which would be for $1295.

If A and B are independent events, it is a simple matter to compute the probability that they occur together, with the use of a theorem called the *Multiplication Rule*.

Theorem 3.2.1. (*The Multiplication Rule*) *If A and B are independent events, then*
$$P(A \cap B) = P(A) \cdot P(B).$$

Informally, the Multiplication Rule states that we can find the probability that two successive independent events occur by multiplying the probability of the first by the probability of the second. Mathematical induction (Appendix B) can be used to extend this rule to any number of independent events: the probability of a sequence of n independent events is simply the product of the n probabilities of the individual events.

Example 3.2.5. We can now use the Multiplication Rule together with the Complement Rule to solve the question posed by the Chevalier de Méré: Why does it seem like the probability of rolling at least one 6 in four tosses of a fair die is slightly more than 50% and the probability of rolling at least one 12 in 24 rolls of 2d6 is slightly less than 50%? In other words, can we develop a theoretical explanation for his observed results?

The probability of rolling at least one 6, by the complement rule, is 1 $P(\text{No 6s})$. The probability of *not* rolling a 6 in a single roll is $\frac{5}{6}$. Since the successive rolls of the die are independent, we have

$$1 - P(\text{No 6s}) = 1 - P(\text{Not a 6})^4 = 1 - \left(\frac{5}{6}\right)^4 = 1 - \frac{625}{1296} = \frac{671}{1296} \approx .518 >$$

consistent with Gombaud's experience that he won slightly more than half the time with this wager.

Similarly, the probability of not rolling a 12 in one roll is $\frac{35}{36}$, and the probability of at least one 12 is $1 - P(\text{No 12s})$, or

$$1 - \left(\frac{35}{36}\right)^{24} \approx .491 < \frac{1}{2},$$

which is also what Gombaud reported.

Note that, by extension, the probability of rolling at least one 6 in n rolls is

$$1 - \left(\frac{5}{6}\right)^n,$$

which approaches but never equals 1 as n approaches ∞.

It is worth noting that the two probabilities in this example are so close to $\frac{1}{2}$ as to cast some doubt on whether Gombaud really gambled at these games long enough to observe the deviation. Since he played a major role in opening up a rich field of mathematics, this discrepancy need not concern us. The story is a good illustration of the difference between theoretical and experimental probability, even if the details may not be completely accurate.

Craps

Craps is a popular casino game played with two six-sided dice. A gam begins when the shooter rolls the two dice, which is called the *come-out* roll. If the come-out roll is 7 or 11, this is an immediate win; if 2, 3, or 12, an immediate loss.

If the come-out roll is any other number, that number becomes the *point* The shooter then continues rolling until he either rolls the point again, or rolls a 7. All other rolls are disregarded (for the purposes of resolving this main bet; there are several other betting options that can be chosen on individual rolls). The shooter wins if he re-rolls the point before a 7, and loses if he rolls a 7 first.

There is a large collection of wagers available to a craps bettor. Figure 3.1 depicts all of the bets available on a standard craps layout. A full-size craps table would include a second betting field for come, pass, and field bets, placed symmetrically to the right of the illustration to duplicate the betting options at the other end of the table.

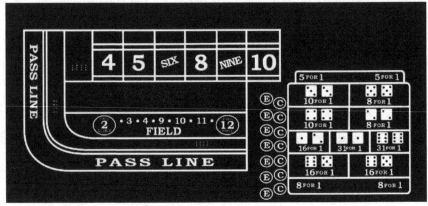

FIGURE 3.1: Craps layout [9]

Example 3.2.6. We shall focus first on the *pass* and *don't pass* bets. A pass bet is a bet that the shooter will win, and a don't pass bet is a bet that the shooter will lose. For the pass bet, there are six ways to roll a 7 and two ways to roll an 11 on the come-out roll, so P(Win on the come-out roll) = 8/36. If a point is established, the probability of an eventual win depends on the value of the point. Once the point is known, the only rolls that matter are those resulting in that number or in a 7. For example, if the point is 9, all we look at are the four ways to roll a 9 and the six ways to roll a 7, so there are ten rolls that will resolve the pass/don't pass question. The probabilities of winning with each point are in this table:

Point	Ways to roll	P(Win)
4 or 10	3	3/9
5 or 9	4	4/10
6 or 8	5	5/11

The probability of each number becoming the point is equal to the number of ways to roll that point divided by 36, so, for example, the probability of first establishing the number 4 as a point and then successfully making that point is $(3/36) \cdot (3/9) = 1/36$. For the other points, we have the following:

Point	P(Point rolled)	P(Point wins)	P(Win on this point)
4 or 10	3/36	3/9	1/36
5 or 9	4/36	4/10	2/45
6 or 8	5/36	5/11	25/396

By adding up the seven probabilities so far determined, we find that the probability of winning a pass line bet is approximately

$$p = \frac{8}{36} + 2 \cdot \frac{1}{36} + 2 \cdot \frac{2}{45} + 2 \cdot \frac{25}{396} = \frac{244}{495} \approx .4929.$$

If there were no other rules, it would follow that the probability of winning a don't pass bet would be $q = 1 - p \approx .5071$, which is greater than 50%—and this is not in the casino's interests, for the gambler would have an advantage.

To prevent this, most casinos have adopted a "Bar 12" rule for the come-out roll, which means that if the shooter rolls a 12 on the come-out roll, pass bets lose but don't pass bets push (tie) instead of winning. With this removed, the probability of winning on a don't pass bet is $p = .4792$. Some casinos bar the 2 rather than the 12; since 2 and 12 are equally likely rolls, the probabilities are the same.

A standard craps layout also includes spaces labeled "Come" and "Don't Come." These wagers act and are paid exactly like Pass and Don't Pass, but may be made by a bettor at any time, without waiting for the previous point to be resolved. When a Come or Don't Come bet is made, the next roll of the dice functions as a come-out roll. If the bet is not resolved immediately by the roll of a 2, 3, 7, 11, or 12, then the chips are moved to a numbered space on the layout just above the "Come" space bearing the point rolled, where the bet can be monitored by casino personnel until it is resolved.

If Pass/Come and Don't Pass/Don't Come were the only bets available at craps, there would be a lot of downtime without much wagering, as it can potentially take many rolls before the pass/don't pass question is resolved. To maintain the level of excitement and encourage additional wagering, a craps layout has many more betting options available to players. One class of craps wagers is *one-roll* bets, which are simply bets on what the next roll of the dice will be and thus are quickly resolved.

Example 3.2.7. A very simple one-roll bet is the *Eleven* bet, which wins if the next roll is an 11 and loses otherwise. There are two ways to roll an 11—6-5 and 5-6—so the probability of winning the Eleven bet is 2/36, or 1/18.

In practice, a roll of 11 is frequently announced with a shout of "Yo!"—this is short for "yo-leven," which is used to distinguish "eleven" from the similar-sounding "seven."

It should be noted that the craps layout depicted in Figure 3.1 gives the payoff for the Eleven bet as "15 for 1," which means that a winning bettor's wager is included as part of the payoff, so that the effective payoff is only 14 to 1. "15 for 1" sounds like a bigger prize than the identical payoff of "14 to 1," which may explain this word choice.

Example 3.2.8. The multiple "C" and "E" spots located to the left of the center betting section in Figure 3.1 stand for Crap and Eleven, and are placed on the felt in such a way that they point at players who gather at the table. By using these betting spots, dice dealers can more easily keep track of which bets belong to which bettor. C and E are used for making the *Craps & Eleven* one-roll bet, which combines the "Any Craps" (2, 3, or 12) bet with the Eleven bet. Any Craps pays off at 8 for 1 if the next roll is 2, 3, or 12; these are the numbers that lose for the Pass better on the come-out. The Eleven bet pays off at 15 for 1 if the next roll is an 11.

In practice, a player's C&E wager is split evenly between the C and E spaces; if this is not possible, as for example with a $5 wager, the chip or chips are placed between the circles, and any fractional payoff is rounded down, in the casino's favor.

The probability of winning the C&E bet can be calculated by adding up the number of ways to roll a 2, a 3, an 11, or a 12, and is simply

$$p = \frac{1+2+2+1}{36} = \frac{6}{36} \approx .1667.$$

Sports Betting

In the USA, a 1992 law, the Professional and Amateur Sports Protection Act, restricts wagering on sporting events to only four states: Delaware, Montana, Nevada, and Oregon. These four states had all legalized sports betting before the passage of the law. Nevada is the only state currently offering the chance to bet on sports directly; the other three states simply have sports-themed lotteries. In 2012, New Jersey announced a plan to introduce sports betting into its casinos and racetracks in defiance of the law.

Example 3.2.9. At 15 casinos in metropolitan Las Vegas operated by Station Casinos, it is possible to wager on sporting events through a daily *parlay card* In playing a parlay card, you place a bet (minimum bet $2) and pick the winner, against a point spread, of 3–10 games being played on a given day. If *all* of your selections win, you are paid off according to Table 3.1.

Once again, the payoffs quoted here are in the form "*x for* 1" rather than "*x to* 1," as we saw with the craps table in Figure 3.1, and as is also seen on some video blackjack machines. If you win, your wager is returned as part of your payoff—for example, if you bet $10 on a five-team parlay card and pick all five games correctly, you will be paid $230, not $240. One reason for quoting odds in this fashion is to make the payoff seem bigger; another reason is that there can be a considerable time interval between when the bet is placed and when the games are completed and the win is verified, and it makes no sense for the casino not to collect your wager when you place the bet.

Since each game can reasonably be considered to be independent of all the others, computing the probability of winning is easy. The casino sets the

TABLE 3.1: Station Casinos' sports parlay card payoffs

Wins	Payoff
3 for 3	6 for 1
4 for 4	12 for 1
5 for 5	23 for 1
6 for 6	45 for 1
7 for 7	80 for 1
8 for 8	160 for 1
9 for 9	320 for 1
10 for 10	800 for 1

point spreads in an effort to encourage equal action on each team in each game, so we will begin by assuming you have a 50% chance of picking each winner correctly. With that probability and the assumption of independence, the probability of winning an n-team parlay is simply $\left(\frac{1}{2}\right)^n$. If you are an expert in predicting winners and can pick the right team 55% of the time, your chance of winning rises to $(.55)^n$.

Conditional Probability

If the events A and B are not independent, we will need to generalize Theorem 3.2.1 to handle the new situation. This generalization requires the idea of *conditional probability*. We begin with an example.

Example 3.2.10. If we draw one card from a standard deck, the probability that it is a king is $\frac{4}{52} = \frac{1}{13}$. If, however, we are told that the card is a face card, the probability that it's a king is $\frac{4}{12} = \frac{1}{3}$—that is, additional information has changed the probability of our event by allowing us to restrict the sample space. If we denote the events "The card is a king" by K and "The card is a face card" by F, this last result is written $P(K|F) = \frac{1}{3}$ and read as "the (conditional) probability of K given F is $\frac{1}{3}$."

The fundamental idea here is that more information can change probabilities. If we know that the event A has occurred and we're interested in the event B, we are now not looking for $P(B)$, but $P(B \cap A)$, because only the part of B that overlaps with A is possible. With that in mind, we have the following formula for conditional probability:

Definition 3.2.2. The *conditional probability* of B given A is

$$P(B|A) = \frac{P(B \cap A)}{P(A)}.$$

This formula divides the probability of the intersection of the two events by the probability of the event that we know has already occurred. Note that if A and B are independent, we immediately have $P(B|A) = P(B)$, since then $P(B \cap A) = P(A) \cdot P(B)$. This is one case where more information—in this case, the knowledge that A has occurred—does not change the probability of B occurring.

Example 3.2.11. Suppose that we draw one card from a standard deck. Let A be the event that the card is an ace and D be the event that the card is a diamond. We have $P(A) = 1/13$, and the event $A \cap D$ is the event that the card is the A$\diamondsuit$. If we are told that the card is a diamond, the probability tha the card is also an ace is

$$P(A|D) = \frac{P(A \cap D)}{P(D)} = \frac{\frac{1}{52}}{\frac{1}{4}} = \frac{4}{52} = \frac{1}{13},$$

which confirms that the events A and D are independent. Knowing the suit of a drawn card gives us no new information about its rank.

Example 3.2.12. In craps, the probability of rolling an 11 is $\frac{1}{18}$. Suppose that, on the come-out roll, one of the dice leaves the table and cannot be seen, but the other die is a 5. What is the probability that the roll is an 11?

Let A be the event {The first die is a 5} and B be the event {The sum is 11}. We seek $P(B|A)$. $P(A) = \frac{1}{6}$, and $P(B \cap A) = \frac{1}{36}$, since the compound event $B \cap A$ can be described as "the roll is a 5 on the first die (the one we can see) and a 6 on the second die." It follows that

$$P(B|A) = \frac{P(B \cap A)}{P(A)} = \frac{\frac{1}{36}}{\frac{1}{6}} = \frac{1}{6}.$$

We see that knowing that one die is a 5 has improved the chance of rolling an 11 from $\frac{1}{18}$ to $\frac{1}{6}$.

It should be noted that, in actual casino play, this roll would be declared void, since a die left the table.

Example 3.2.13. Recall from Example 2.4.12 that the probability of being dealt pocket aces at Texas Hold'em is $\dfrac{1}{221}$. If you have indeed been dealt two aces, what is the probability that the player to your immediate left also holds two aces in the hole?

We can answer this question by restricting the sample space. With your cards removed from consideration, there are now $_{50}C_2 = 1225$ possible two-card combinations available for the next player. Of these, only one is a pair of aces (since only two aces remain among the unknown cards), so the probability of a second pair of pocket aces, given that one pair has already been dealt, is

$$P(2 \text{ aces}|2 \text{ aces}) = \frac{1}{1225}$$

—about five and a half times less, which gives more support to the idea that you should bet your holdings strongly.

As with the addition rules, we can state a second, more general, version of the Multiplication Rule that applies to any two events—independent or not—and reduces to the first rule when the events are independent. This more general rule simply incorporates the conditional probability of B given A, since we are looking for the probability that both occur.

Theorem 3.2.2. *(The General Multiplication Rule) For any two events A and B, we have*

$$P(A \cap B) = P(A) \cdot P(B|A).$$

Proof. This result follows from the fact that $P(A \cap B) = P(B \cap A)$ and from Definition 3.2.2.

Table 3.4 at the end of this chapter (page 71) collects the most important probability formulas.

Bingo

As an example of this more general rule, we consider *bingo*. Bingo is related to lotteries and keno, and is a game familiar to many from its frequent appearances in charitable fundraisers. Foxwoods in Mashantucket, Connecticut, one of North America's largest casinos, got its start as a bingo hall on the Mashantucket Pequot reservation.

Bingo is a simple game to play: players receive cards on which the word "BINGO" is printed above a 5 × 5 grid of numbers, as in Figure 3.2. The five squares under the B are filled with numbers in the range 1–15, the I squares from 16 to 20, the N squares from 31 to 45, the G squares from 46 to 60, and the O squares from 61 to 75. The center square of the grid, under the N, is traditionally a "free" square and contains no number.

B	I	N	G	O
10	30	41	51	68
6	22	44	58	64
4	26	Free	56	65
15	24	43	48	75
5	17	33	50	68

FIGURE 3.2: Bingo card

Numbers in the range from 1 to 75 are drawn, either from a cage or blower full of ping-pong balls as in keno, or electronically. As numbers that appear on the players' cards are drawn, the players cover them in some fashion. The free space is considered covered automatically. Depending on the rules of the game, the winner is the first player to complete a line of five covered spaces horizontally, vertically, or diagonally, including the center free space; cover all

24 numbers (the "Cover All" game); or make other significant patterns, such as crosses or Xs, on their card. A winning player traditionally shouts "Bingo!" to indicate their success and have their win certified so that they may claim their prize.

Casinos have been known to offer bingo as a *loss leader*: a game on which they are prepared to lose money in order to bring people into the building. Since casino bingo is typically scheduled in discrete sessions, often one starting every other hour, rather than being continuously played, the thought behind this strategy is that bingo players will venture into other areas of the casino and gamble there—in less favorable games—in between sessions. The reason that bingo may be a losing proposition for a casino is that there is always a winner in every game, no matter how few people might be playing, and if the amount of the top prize is not covered by player wagers, the game still goes on.

Example 3.2.14. How many possible bingo cards are there?

Order matters when laying out a bingo card—if the first five numb drawn are B-13, I-28, N-41, G-55, and O-72, and all five appear on your card, you are only a winner if they are all in the same row or if all except 41 (the N number) fall on a diagonal. For the B, I, G, and O columns, there are $_{15}P_5 = 360,360$ possible choices of the numbers. In the N column, the free space means that there are only 4 numbers to be picked, and so there are $_{15}P_4 = 32,760$ ways to fill it in. The total number of cards is therefore

$$(360,360)^4 \cdot 32,760 = 552,446,474,061,128,648,601,600,000$$

—or about 552 septillion.

Example 3.2.15. Since the chance of winning at bingo depends both on the numbers on one's own card and the numbers on the other players' cards, it can be challenging to compute the probability of winning. We shall consider a simple question: What is the probability of covering all 24 of your numbers after exactly N numbers have been called?

As with keno, we need to account for all N numbers drawn. Twenty-four of these must be the numbers that appear on your card, and one of them must be the last number (otherwise you would win in fewer than N numbers). It follows that the first $N - 1$ numbers drawn must contain 23 of yours, and then that, given the occurrence of this event, the last draw must be your 24th number. Define the following events:

- A = Draw 23 numbers in exactly $N - 1$ bingo balls.

- B = Draw your 24th number on the Nth ball.

Then P(Draw 24 numbers in exactly N balls) is $P(A \cap B) = P(A) \cdot P(B$

For $P(A)$, there are $_{24}C_{23} = 24$ ways to pick those 23 numbers. The remaining $(N - 1) - 23 = N - 24$ numbers must be drawn from among the

51 numbers that are not on your card; those may be chosen in $_{51}C_{N-24}$ ways. There is a total of $_{75}C_{N-1}$ ways to draw the first $N-1$ numbers. Assembling these factors gives

$$P(A) = P(\text{Draw 23 numbers in } N-1 \text{ bingo balls}) = \frac{_{24}C_{23} \cdot {}_{51}C_{N-24}}{_{75}C_{N-1}}$$

Having accounted for this event, we now consider the probability of drawing that last number on the Nth bingo ball, given that 23 numbers have already been drawn. When the last number is drawn, there are $75 - (N-1) = 76 - N$ bingo balls remaining, and only one of them bears the lone uncovered number on your card. The probability of matching that number is therefore $P(B|A) = 1/(76 - N)$.

In the N draws, then, we have the following probability:

$$P(\text{Win in exactly } N \text{ draws}) = \left(\frac{_{24}C_{23} \cdot {}_{51}C_{N-24}}{_{75}C_{N-1}} \right) \cdot \left(\frac{1}{76 - N} \right).$$

We shall illustrate the use of this formula by considering the extreme cases. If $N = 24$, the formula simplifies to

$$P(\text{Win in 24 draws}) = \left(\frac{24}{_{75}C_{23}} \right) \cdot \left(\frac{1}{52} \right) = \frac{24 \cdot 23! \cdot 52!}{75! \cdot 52} = \frac{24! \cdot 51!}{75!} = \frac{1}{_{75}C}$$

as we would expect, since we're asking for the probability of drawing precisely your 24 numbers in 24 draws.

If $N = 75$, then we have

$$P(\text{Win in 75 draws}) = \frac{24}{75} = .3200,$$

which is just the probability of choosing one of the numbers on your card from the full range of 75 numbers. This may be interpreted as choosing one number *not* to be drawn from 75, and thinking of that number as the last number left in the cage after 74 numbers have been drawn.

In general, the expression for this probability may be rewritten as

$$P(\text{Win in } N \text{ draws}) = \frac{24 \cdot 51! \cdot (N-1)!}{(N-24)! \cdot 75!}.$$

Table 3.2 contains this probability for various values of N.

At the same time, we might be interested in the probability of covering a complete card in N *or fewer* draws. Computing that probability is a simple matter of adding up the individual mutually exclusive probabilities $P(\text{Cover all in } x \text{ draws})$ for $24 \le x \le N$:

$$P(\text{Cover all in} \le N \text{ draws}) = \sum_{k=24}^{N} \frac{24 \cdot 51! \cdot (k-1)!}{(k-24)! \cdot 75!}.$$

TABLE 3.2: Bingo probabilities for Cover All

N	P(Cover All in Exactly N Draws)
24	3.8792×10^{-20}
25	9.3100×10^{-19}
30	1.8427×10^{-14}
35	1.1098×10^{-11}
40	1.4629×10^{-9}
45	7.8073×10^{-8}
50	2.2632×10^{-6}
55	4.2125×10^{-5}
60	5.5941×10^{-4}
65	5.6916×10^{-3}
70	4.6664×10^{-2}
75	$.3200$

Some of these cumulative probabilities are collected in Table 3.3.

TABLE 3.3: Cumulative bingo probabilities for Cover All

N	P(Cover All in $\leq N$ Draws)
25	9.6979×10^{-19}
35	1.6185×10^{-11}
45	1.4639×10^{-7}
55	9.6537×10^{-5}
65	1.5415×10^{-2}
75	1

In practice, many cards are in play in a single round of bingo: there are surely multiple players, and it is not uncommon for bingo players to play many cards in one game. Calculating the probability of holding the *first* card to win a Cover All game is extremely tricky because two cards are generally not independent; they may share one or more numbers. A rough approximation of the chance of any one card being the first to have all of its numbers called can best be done by repeated simulation—see [71] for the details.

Punchboards

The *punchboard* was a popular, though often crooked, gambling device frequently found in American retail stores as a trade stimulator from the late 1700s through the middle of the 20th century [60]. A common form of punchboard consists of a thick piece of cardboard with anywhere from 20 to 10,000 small holes, called *spots*, cut into it and covered on both sides by paper or thin metal foil. The holes each contain a small slip of paper printed with a number. One such board is shown in Figure 3.3.

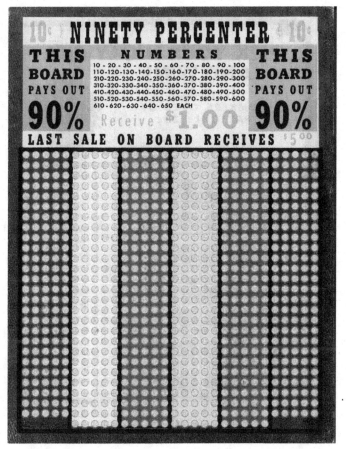

FIGURE 3.3: The Ninety Percenter punchboard

Upon payment of an entry fee, which typically ranged between 5¢ and $ player would punch out a circle to reveal their number. Certain of these numbers corresponded to prizes of cash or merchandise, which could be redeemed on the spot. Proprietors would purchase punchboards from local dealers for an amount less than the difference between the cost of all of the spots and the sum of all prizes; this is how a merchant could make money from a punchboard in his or her store.

An alternate punchboard game replaced the numbers with girls' names, as in Figure 3.4. Each spot sold corresponded to one name; the person buying each spot signed the punchboard beside their chosen name—perhaps the name of a wife or girlfriend—and then punched out the spot to reveal the price of that name. Once all of the spots were sold, a seal on the punchboard was broken to reveal the prize-winning name.

FIGURE 3.4: The Lucky Girl punchboard

Some punchboards, called *keyed* punchboards, were sold together with a list or diagram of the locations of the winning numbers for the use of the proprietor, and while this could easily be explained as a convenience, it was just as easy for an unscrupulous operator to either steer his preferred customers to the winning numbers or punch out all of the biggest winners before putting the board on display so that the top prizes could not be won, thus increasing his profit. Alternately, an associate of the punchboard dealer could obtain a key from him, take it around to neighborhoods where punchboards were available for play, and punch out winning spots in what appeared to be a run of pure luck. The spread of crooked punchboards led to a rapid decline in their popularity; however, the punchboard features of player involvement in a game and an instant decision make them a forerunner of the scratch-off instant lottery tickets that remain popular today.

Example 3.2.16. Suppose that you're playing the Lucky Girl punchboard in Figure 3.4, which has 35 spots, and that the top prize is $5. If you are the first to purchase a chance, your probability of winning is $1/35$. What is your chance of winning (that is, of picking the winning name) if you are the second player?

To win on the second punch, the first punch must have chosen a losing name. The probability of losing on the first punch is

$$P(A_1) = \frac{34}{35},$$

and the probability of then winning on the second punch is

$$P(A_2|A_1) = \frac{1}{34},$$

since we can compute this conditional probability by restricting the sample space to the 34 spots left after the initial loser. These events must occur in sequence, which gives

$$P(\text{Lose on first punch and Win on second punch}) = P(A_1 \cap A_2).$$

By the General Multiplication Rule , this is

$$P(A_1 \cap A_2) = P(A_2|A_1) \cdot P(A_1) = \frac{1}{34} \cdot \frac{34}{35} = \frac{1}{35},$$

which is the same as the probability of winning on the first punch.

This assumes, of course, that the punchboard operator is offering a fair game, which was often not the case.

Repeating this calculation with different numbers of previous losing punches reveals an interesting and possibly counterintuitive result: *The probability of winning any specified prize is independent of the number of previous punches.* There is no benefit in going first, nor in waiting until the end; advantages and disadvantages cancel out at any place in the line.

Example 3.2.17. To illustrate this independence, we consider a generalization: For a punchboard with n spots and a single prize, find the probability that the jackpot will be won by the player picking the last spot.

We need only compute the probability that the first $n-1$ spots chosen do not win the jackpot, for then the jackpot number must be under the last remaining spot. The probability of losing on the first punch is

$$p_1 = \frac{n-1}{n}.$$

Given that the first punch loses, the probability that the second punch also loses is

$$p_2 = \frac{n-2}{n-1}.$$

We continue in this fashion:

$$p_k = P(\text{Lose on punch } \#k \mid \text{All previous punches lose}) = \frac{n-k}{n-(k-1)}.$$

Multiplying these probabilities together for $k = 1, 2, \ldots, n-1$ yields

$$P(\text{Win on punch } \#n) = \frac{n-1}{n} \cdot \frac{n-2}{n-1} \cdot \frac{n-3}{n-2} \cdots \frac{3}{4} \cdot \frac{2}{3} \cdot \frac{1}{2} = \frac{1}{n},$$

since all of the intermediate numerators and denominators cancel.

This is the same chance of winning that you have if you pick first or second—or anywhere else in line.

Example 3.2.18. Another example of this independence is a contest like some run by radio stations. Eight keys, one of which starts a car, are placed in a bowl, and eight contestants are each invited to choose a key and see if it starts the car. If the first key chosen does not start the car, the second contestant chooses, and so on until the car starts and is awarded to the holder of the lucky key. People frequently think that it is better to choose first, or second, or somewhere else in the line—but the analysis above can be used to show that every contestant has an equal chance to win.

Example 3.2.19. The *Ninety Percenter* punchboard (Figure 3.3) claims just that: a 90% payout of all money invested by players. The punchboard has 800 spots, which are sold for 10¢ each. Players punching out any multiple of 10 between 10 and 650 receive a $1 prize, and the last spot on the board carries a $5 bonus. If the claim is accurate and all of the winning numbers appear at least once each, how many multiples of 10 are repeated on this punchboard?

If all 800 spots are sold, the proprietor takes in $80. If the claim of a 90% payback rate is correct, then $72 must be returned in prizes. Subtracting the $5 bonus for the last spot leaves $67, so two extra winning spots must be included—either two numbers repeated twice each or one number repeated three times.

If we make the reasonable assumption that each winning number appears only once, then the total prize money paid out is $70, and the return percentage is 87.5%—which only becomes 90% through generous rounding.

Dice Games

Example 3.2.20. Here's a problem posed by noted gambling mathematician Peter Griffin in the magazine *Casino & Sports* and reprinted in [22]:

> *You have your choice of betting that a pair of sixes will appear on a roll of 2d6 before back-to-back totals of seven appear, or taking the opposite position: that two consecutive sums of seven will be rolled before a 12. Which side should you take?*

At first glance, this looks like an even proposition: the probability of rolling a 7 is $\frac{1}{6}$, and the probability of rolling two 7s in succession is thus $\frac{1}{36}$—the same as the probability of rolling a 12 on 2d6. Griffin reported that "an enterprising pair from New York was offering this proposition in Atlantic City," but that they were backing the less favorable outcome. This suggests that the proposed game is not as even as it may first appear. Which is the favored side?

Successive rolls of the dice are independent, and so the Multiplication Rule is useful in solving the problem. Denote by A the event that a 12 is rolled before two consecutive 7s. If $P(A) > .5$, this side has the advantage; if $P(A) < .5$, the other side is favored. A is itself the union of infinitely many mutually exclusive events:

- $A_1 = $ A 12 is rolled on the first throw. $P(A_1) = \frac{1}{36}$.

- $A_2 = $ A 12 is not rolled on the first throw, but is then rolled on the second throw. For convenience, we divide A_2 into two subcases: the case where a 7 is rolled on the first toss and then a 12 comes up on the second, and the case where the first roll is something other than a 7 or 12 (a "neutral" roll) and a 12 is rolled on the second toss. It then follows that

$$P(A_2) = \frac{1}{6} \cdot \frac{1}{36} + \frac{29}{36} \cdot \frac{1}{36} = \frac{35}{1296}.$$

- $A_3 = $ The first two tosses produce neither a 12 nor two straight 7s, a 12 is then thrown on the third roll. We have

$$P(A_3) = \left[\left(\frac{35}{36} \right)^2 - \frac{1}{36} \right] \cdot \frac{1}{36} = \frac{1189}{46,656}.$$

- And so on. The probabilities become smaller as n, the number of rolls, increases, but $P(A_n)$ is never 0.

Let $P(A)$ be denoted by p. Since the events A_n are mutually exclusive, the Addition Rule gives the infinite series

$$p = \sum_{n=0}^{\infty} P(A_n) = \underbrace{\frac{1}{36}}_{P(A_1)} + \underbrace{\frac{1}{6} \cdot \frac{1}{36} + \frac{29}{36} \cdot \frac{1}{36}}_{P(A_2)} + \underbrace{\frac{35^2 - 36}{36^2} \cdot \frac{1}{36}}_{P(A_3)} + \ldots$$

Some algebraic gymnastics will allow us to convert this infinite sum of probabilities into something easily managed. If the first roll is a 12, the game is over, and similarly if the first roll is a 7 but the second is a 12. If the first toss is neutral (probability $\frac{29}{36}$), then we can think of the game as starting on the second roll, with $P(A) = p$. Finally, if the first toss is a 7 but the second is neutral, we're back at the beginning again, and the probability p reappears as a factor. Taken together, we have

$$p = \left(\frac{1}{36} \right) + \left(\frac{1}{6} \cdot \frac{1}{36} \right) + \left(\frac{29}{36} \cdot p \right) + \left(\frac{1}{6} \cdot \frac{29}{36} \cdot p \right).$$

This finite sum is a linear equation in p and hence is easily solved. Collecting terms on the right gives

$$p = \frac{7}{216} + \frac{203}{216} \cdot p,$$

from which we find that

$$p = \frac{7}{13} \approx .538,$$

and so it is more likely that a 12 will be rolled before a pair of consecutive 7s.

Example 3.2.21. The casino game *Die Rich* was introduced at Las Vegas' Luxor casino in 2007. The game is played with one standard six-sided die, as follows:

- On the first roll, 6 wins and 1 loses. If a 6 is rolled, the payoff is 1 to 1.

- If neither a 1 nor a 6 is rolled, the number rolled becomes the "point," and the shooter rolls up to three more times. If a 1 comes up on any of these three rolls, the shooter loses immediately.

- If the point comes up on the first or last of these rolls, the payoff is 2 to 1.

- If the point comes up on the second roll, the payoff is 1 to 1.

One advantage of this game over traditional craps is that a round of Die Rich will always be completed in no more than four rolls. In traditional craps, it theoretically can, and frequently does, take a large number of rolls to resolve a game.

There are a number of paths we can take from the first roll to the resolution of the bet. We define the following events, where $n = 1, \ldots, 6$:

- A_n: The first roll is the number n. Note that the rules of the game are that if the first roll is a 1 or 6, there are no further rolls.

- B_n: The second roll is the number n.

- C_n: The third roll is the number n.

- D_n: The fourth roll is the number n.

If we denote the point number by m, $m = 2, \ldots, 5$, the following events constitute a win for the shooter where, for example, B_m^* denotes the event that the second roll is neither m nor 1, and thus the game continues.:

- A_6: Roll a 6 on the first roll.

- $B_m | A_m$: Roll the point on the first roll and then again on the second roll.

- $C_m|(B_m^*|A_m)$: Roll the point on the first roll, roll neither 1 nor the point on the second, and roll the point on the third.

- $D_m|(C_m^*|(B_m^*|A_m))$: Roll the point on the first roll, complete two non-deciding rolls, and re-roll the point on the fourth roll.

The events arising from successive rolls of the die are independent, and so we have the following winning probabilities:

Event	P(Event)	Payoff
A_6	$\frac{1}{6}$	1
$B_m|A_m$	$\frac{1}{6} \cdot \frac{4}{6} = \frac{1}{9}$	2
$C_m|(B_m^*|A_m)$	$\frac{1}{6} \cdot \frac{4}{6} \cdot \frac{4}{6} = \frac{2}{27}$	1
$D_m|(C_m^*|(B_m^*|A_m))$	$\frac{1}{6} \cdot \frac{4}{6} \cdot \frac{4}{6} \cdot \frac{4}{6} = \frac{4}{81}$	2

Note that $P(A_m) = \frac{4}{6}$, since there are four possible point numbers, and all are equally likely. Adding up the second column gives $P(\text{Win}) = \dfrac{65}{162} \approx .4012$ and so $P(\text{Lose}) = \dfrac{97}{162} \approx .5988$.

Table 3.4 collects some of the more useful probability rules that we have discussed so far. In referring to this table, note that the First Addition Rule applies only to mutually exclusive events, while the Second Addition Rule applies to all events. Similarly, the first Multiplication Rule in the table requires that the events A and B be independent; the General Multiplication Rule applies in all cases.

TABLE 3.4: Common probability formulas

Complement Rule	$P(A') = 1 - P(A).$
Subset Rule	If $B \subset A$, then $P(B) \le P(A).$
Permutations	$_nP_r = \dfrac{n!}{(n-r)!}.$
Combinations	$_nC_r = \dfrac{n!}{r! \cdot (n-r)!}.$
First Addition Rule (mutually exclusive events)	$P(A \cup B) = P(A) + P(B).$
Second Addition Rule	$P(A \cup B) = P(A) + P(B) - P(A \cap B)$
Multiplication Rule (independent events)	$P(A \cap B) = P(A) \cdot P(B).$
Conditional Probability	$P(B|A) = \dfrac{P(A \cap B)}{P(A)}.$
General Multiplication Rule	$P(A \cap B) = P(A) \cdot P(B|A).$

3.3 EXERCISES

Answers begin on page 254.

3.1. Video poker (see Section 4.2) is an electronic version of five-card draw poker. When the option to discard some cards and draw replacements is included, the probability of a royal flush is roughly 1 in 40,000. Use this estimate to calculate the probability of *not* drawing a royal flush in 40,000 video poker hands.

3.2. In Texas Hold'em, suppose that your hole cards are an ace and a 2. Find the probability that the player to your immediate left has been dealt pocket aces.

3.3. Two bingo cards with different sets of 24 numbers are called *disjoint* cards with one or more numbers in common are said to *overlap*. Given the layout of a card, it is possible to construct three disjoint cards—the 72 numbers on the three cards include all 75 numbers except three of the numbers from 31 to 45, due to the free space under the N. In a Cover All game with these three cards, find the probability that there will be a winner by the time that the 65th number is drawn.

3.4. Find the probability that two randomly selected bingo cards will have no numbers in common.

3.5. Bingo cards are typically sold in packages of 3000. Use the probability calculated in Exercise 3.4 to find the probability that a package will contain two or more disjoint cards.

3.6. If you make the Outside Eagle's Edge keno bet (Example 2.4.24), you lose your wager if exactly 6, 7, 8, or 9 edge numbers are drawn. Find the probability of this event.

3.7. *Mini-craps* is a version of craps that is played on a smaller table and frequently offers simpler betting options. One version of mini-craps removes the four line bets: pass, don't pass, come, and don't come. These are replaced with the simpler "Over 7" and "Under 7" bets, where the gambler simply wagers on the value of the next roll of the dice. These wagers pay off at even money. Find the probability of winning the Over 7 bet, which wins if the subsequent roll is 8 or higher.

3.8. Exercise 2.14 describes the $20 keno wager at the Colorado Belle casino in Laughlin, Nevada, that pays off a $100,000 jackpot if at least 15 of a player's 20 numbers are drawn by the casino. What is the probability that a single keno ticket will win the $100,000 prize?

3.9. A *complete bet* at roulette is a collection of bets that cover every possible way, through split, street, corner, and double street bets, to make an inside bet on a single number. The cost of a complete bet depends on the number selected, which affects the number of some of the compound bets that are available. For example, a complete bet on the number 23 includes a single-number bet on 23; four split bets covering 23 with one other adjacent number; one street bet on 22, 23, and 24; four corner bets; and two double street bets. At $1 per bet, this wager would cost $12 altogether.

a. Assuming an American roulette wheel, find the probability of hitting a winning number on a complete bet on 23.

b. Referring to Figure 2.2, how many separate $1 bets must be made to make a complete bet on the number 33? Find the probability of hitting a winning number, on an American roulette wheel, for this bet.

3.10. Find the probability that each number on an American roulette wheel will come up exactly once in a string of 38 consecutive spins.

Spider Craps

Spider craps is a game variation developed by Jacob Engel that uses eight-sided dice rather than six-sided dice. In order to develop a game that will be attractive both to players and to casinos, certain modifications are made that are nonetheless consistent with the rules of the original game. The following exercises explain and explore some of the rules of spider craps.

3.11. On the come-out roll in spider craps, a Pass bet wins instantly if the shooter rolls a 9 or a 15 and loses if the roll is a 2, 3, or 16. Find the probability of winning a Pass bet and the probability of losing a Pass bet on the come-out roll.

3.12. In spider craps, once a point (4-8 or 10-14) has been established on the come-out roll, a Pass line bet wins if the shooter re-rolls that number before rolling a 9. By mimicking the calculations leading to the table on page 56, compute the probability of winning a Pass line bet at spider craps.

3.13. If no rolls are barred on the come-out roll, then a Don't Pass bet at spider craps will have a probability of winning that exceeds .5. Show that barring only the 16 (analogous to standard craps), or the 2 and 16 together, still results in a bet with an advantage for the player.

3.14. Show that if both 2 and 3 are barred on the come-out roll, then t spider craps Don't Pass bet has a probability of winning that is under 50%. This is the rule in force in spider craps.

Probability Distributions and Expectation

4.1 RANDOM VARIABLES

Definition 4.1.1. A *random variable* (RV for short) is an unknown quantity X whose value is determined by a chance process.

This is another definition that, on its face, isn't terribly useful—indeed, this comes perilously close to using the words "random" and "variable" in its own definition. Once again, a sequence of examples will illustrate this important idea far better than a formal definition.

Example 4.1.1. Roll 2d6 and let X denote their sum. X then takes on a value in the set {2, 3, 4, 5, 6, 7, 8, 9, 10, 11, 12}.

Example 4.1.2. In a five-card poker hand, let X count the number of aces it contains. $X \in \{0, 1, 2, 3, 4\}$.

Example 4.1.3. In a hand of blackjack, let X denote the sum of the first two cards, counting the first ace as 11. Here, $X \in \{3, 4, 5, 6, \ldots, 19, 20, 21\}$ (A hand containing two aces would be counted here as 12, not 2 or 22.)

Example 4.1.4. Let X be the number spun on a European roulette wheel. Then $X \in \{0, 1, 2, 3, \ldots, 35, 36\}$, a set with 37 elements.

Definition 4.1.2. A *probability distribution* for a random variable X is a list of the possible values of X, together with their associated probabilities.

Example 4.1.5. Suppose we roll 2d6 and let X denote the sum of the numbers rolled. The possible sums are illustrated in Table 4.1, and the corresponding probability distribution for X is compiled in Table 4.2.

TABLE 4.1: Sample space of outcomes when rolling 2d6

X					
2	3	4	5	6	7
3	4	5	6	7	8
4	5	6	7	8	9
5	6	7	8	9	10
6	7	8	9	10	11
7	8	9	10	11	12

TABLE 4.2: Probability distribution when rolling 2d6

x	2	3	4	5	6	7	8	9	10	11	12
$P(X = x)$	$\frac{1}{36}$	$\frac{2}{36}$	$\frac{3}{36}$	$\frac{4}{36}$	$\frac{5}{36}$	$\frac{6}{36}$	$\frac{5}{36}$	$\frac{4}{36}$	$\frac{3}{36}$	$\frac{2}{36}$	$\frac{1}{36}$

Axioms 2 and 3 require that the sum of the probabilities in Table 4.2, and in any probability distribution, is 1.

To ensure that their craps tables are profitable, casinos depend on their dice following this probability distribution. In order to guarantee that the real dice conform to these theoretical probabilities, casino dice are manufactured to exacting specifications and are perfect cubes to within 1/10,000 of an inch on each side. The spots on the dice are made of a different-colored plastic with the same density as that used for the dice and are flush with the surface rather then being indented, as is the case with common game dice. Moreover, casino dice have razor-sharp edges and corners, in order to ensure that they bounce and roll in a truly random fashion.

Additionally, casino dice are typically used on the tables for eight hours or less and are then removed from play before they develop small chips or other imperfections that might cause them to deviate from this distribution. Dice so removed are immediately canceled with a circle punched into one face or a hole drilled through two faces to guard against reintroduction of modified dice to the tables, and replaced by new dice. (In New Jersey, dice removed from play must be canceled by drilling.) Depending on casino policy, canceled dice may either be destroyed, given away to players, or offered for sale in the casino's gift shop.

A similarly short lifetime on the casino floor awaits playing cards, which are typically used for one eight-hour shift or less and then replaced before they have a chance to accumulate telltale nicks or marks that might tip off an alert gambler to the imminent arrival of a particularly high or low card. Cards used in baccarat or blackjack are also prone to develop "waves" from being bent back by players peering at the card faces, and these curves can also indicate a rank if a player is paying very close attention. Since this would lead to a departure from the true randomness on which the rules of the games depend, and since playing cards are relatively inexpensive, this is a small investment in ensuring that the games are being played correctly. Large casinos go through thousands of decks every week; the Bellagio in Las Vegas has a standing order with its card manufacturer for 60 cases of cards (8640 decks) per week [39]. As with dice, retired card decks are canceled, either by drilling (see Figure 5.4) or clipping off one corner, before being discarded or sold as souvenirs.

Example 4.1.6. *Door pops* are a relatively unsophisticated type of loaded dice. One die of the pair has a 2 on three sides and a 6 on the other three, and the other has a 5 on all six sides [59, p. 246]. It follows that a pair of door pops can only roll a 7 or 11—and thus will always produce a winner on a come-out roll. It stands to reason that such a pair of dice, if introduced into a craps game, would be very easily discovered. On the other hand, the author has, on more than one occasion, presented a pair of door pops to a class of students and watched some of them closely examine the dice for several minutes before figuring out how they differ from standard dice.

If we define X to be the sum of the numbers showing when a pair of door pops is rolled, we have the following probability distribution for X:

x	7	11
$P(X = x)$	$\frac{1}{2}$	$\frac{1}{2}$

Example 4.1.7. *Chuck-a-luck* is a game played with three dice. In its original form, it is not often seen in casinos anymore, but can still be found in carnivals and, in expanded form, in the casino game *sic bo* (see Example 5.2.12). The three dice are spun in a wire cage. Gamblers bet on the numbers from 1 to 6, and are paid according to how many of their number appear on the dice: the amount wagered is matched for each die showing the selected number, so the payoff is 1 to 1 if one die shows the number, 2 to 1 if two do, and 3 to 1 if all three do.

Let X be the number of 4s that appear when the cage is spun. Since the are six sides on each die, there are $6 \cdot 6 \cdot 6 = 216$ ways that the dice can land. We note that the numbers showing on the three dice are independent random variables. We shall consider each possible value for X individually.

$\boxed{X = 0 :}$ There are five sides on each die that do not show a 4, so there are $5 \cdot 5 \cdot 5 = 125$ ways not to roll any 4s. Accordingly, $P(X = 0) = \dfrac{125}{216}$.

$\boxed{X = 1 :}$ For convenience, we shall imagine that the three dice are different colors: red, green, and blue. (In practice, the cage contains three identical dice.) Since there are five ways on each die not to roll a 4, and one way to roll a 4, we have the following chart of possibilities:

Red	Green	Blue	Count
4	Not 4	Not 4	$1 \cdot 5 \cdot 5 = 25$
Not 4	4	Not 4	$5 \cdot 1 \cdot 5 = 25$
Not 4	Not 4	4	$5 \cdot 5 \cdot 1 = 25$

Adding the last column shows that there are 75 ways to roll exactly one 4, so we conclude that $P(X = 1) = \dfrac{75}{216}$.

$\boxed{X = 2:}$ Once again, we organize our work in a chart:

Red	Green	Blue	Count
4	4	Not 4	$1 \cdot 1 \cdot 5 = 5$
4	Not 4	4	$1 \cdot 5 \cdot 1 = 5$
Not 4	4	4	$5 \cdot 1 \cdot 1 = 5$

Adding the last column shows that there are 15 ways to roll exactly two 4s, so we have $P(X = 2) = \dfrac{15}{216}$.

$\boxed{X = 3:}$ There is only one 4 on each die, so there is only one way to roll three 4s. We conclude that $P(X = 3) = \dfrac{1}{216}$. This probability could also have been calculated by using the complement rule and the three probabilities previously computed.

Combining this information gives us the following probability distribution for X:

x	0	1	2	3
$P(X = x)$	$\frac{125}{216}$	$\frac{75}{216}$	$\frac{15}{216}$	$\frac{1}{216}$

Example 4.1.8. The Meskwaki Casino in Tama, Iowa offers a "Super 20 Special" keno game whose brochure advertises "18-Out-Of-21-Ways To Win," For a minimum bet of \$5, the player picks 20 numbers in the range 1–80 and wins unless 4, 5, or 6 of the numbers selected are matched by the casino's numbers—so indeed 18 out of the 21 possible numbers of matches are winners, although matching 2, 3, 7, or 8 numbers merely gets the player their money back. Let X count the number of matches. Since we need to account for every number drawn by the casino in computing probabilities, we have the following formula for $P(X = x)$, which could be used to compute the numerical probabilities that comprise a distribution:

$$P(X = x) = \frac{(_{20}C_x) \cdot (_{60}C_{20-x})}{_{80}C_{20}}.$$

In the numerator, the factor $_{20}C_x$ corresponds to the number of ways to choose x numbers from the 20 selected by the player. The second factor $_{60}C_{20-x}$, counts the number of ways to choose the remaining $20 - x$ numbers from among the 60 that the player did not choose. The denominator remains $_{80}C_{20}$, the number of possible subsets of 20 keno balls from a set of 80.

What can we conclude from this probability function?

- The probability of losing \$5 by matching 4, 5, or 6 numbers is

$$P(X = 4) + P(X = 5) + P(X = 6) = .6285,$$

  which conforms to our notion that keno is not a very player-friendly game.

- Including the four break-even payoffs, the probability of not profiting from this bet is .9662, which definitely makes this look more like a casino game.

4.2 EXPECTED VALUE

The notion of *expected value* is fundamental to any discussion of random variables and is especially important when those random variables arise from a gambling game. The expected value of a random variable X is, in some sense, an "average" value, or what we might expect in the long run if we were to sample many values of X.

The common notion of "average" corresponds to what mathematicians call the *mean* of a data set: add up all of the numbers and divide by how many numbers there are. For a random variable X, this approach requires some fine-tuning, as there is no guarantee that a small sample of values of X will be representative of the range of possible values. Our interpretation of average will incorporate each possible value of X together with its probability, computing what is in some sense a long-term average over a very large hypothetical sample.

Definition 4.2.1. The *expected value* or *expectation* $E(X)$ of a random variable X is computed by multiplying each possible value for X by its corresponding probability and then adding the resulting products:

$$E(X) = \sum_{x} x \cdot P(X = x).$$

This expression may be interpreted as a standard mathematical mean drawn from an infinitely large random sample. If we were to draw such a sample, we would expect that the *proportion* of sample elements with the value x would be $P(X = x)$; adding up over all values of x gives this formula for $E(X)$.

We may abbreviate $E(X)$ to E when the random variable is clearly understood. The notation $\mu = E(X)$, where μ is the Greek letter mu, is also common, particularly when the expected value appears as a term in another expression.

Example 4.2.1. Let X be the result when a fair d6 is rolled. The expectation is

$$E(X) = \sum_{x=1}^{6} x \cdot P(X = x) = \sum_{x=1}^{6} \frac{x}{6} = \frac{21}{6} = 3.5.$$

The notation used in Definition 4.2.1 does not indicate the limits of the indexing variable x, as we have done with summation in earlier chapters; this is because those values may not be a simple list running from 1 to some When written this way, we should take this sum over *all* possible values of the random variable X, as in the next example.

Example 4.2.2. If we let X be the outcome when $1 is wagered on a roulette corner bet (four numbers), the expression for $E(X)$ would be

$$E(X) = \sum_{x} x \cdot P(X = x),$$

where $x = -1$ or $x = 8$. The resulting expectation would be

$$E(X) = (-1) \cdot \frac{34}{38} + (8) \cdot \frac{4}{38} = -\frac{2}{38}.$$

Definition 4.2.2. If X is a random variable measuring the payoffs from a game, we say that the game is *fair* if $E(X) = 0$.

Example 4.2.3. Suppose you gamble with a friend on the toss of a coin. If heads is tossed, you win $1; if tails is tossed, you pay $1. Since a fair coin can be expected to land heads and tails equally often, the expected value for this game is $E = (1)(\frac{1}{2}) + (-1)(\frac{1}{2}) = 0$. The game is fair.

If a game is fair, then in the long run, we expect to win exactly as much money as we lose, and thus, aside from any possible entertainment derived from playing, we expect no gain. This is often summarized in the following maxim:

> *If a game is fair, don't bother to play.*
> *If a game is unfair, make sure it's unfair in your favor.*

Failure to heed this maxim, of course, is responsible for the ongoing success of the gambling industry, for, as we shall see, games which are unfair and favor the gambler are rare.

Example 4.2.4. In Example 2.4.6, we looked at the Baldini's promotion offering a million-dollar payoff for a dealt sequential royal flush at video poker. Considering only the million-dollar prize, what is the expectation of this game?

There is no cost to play this game, so this is one of those rare cases where the advantage lies with the player. However, having the advantage doesn't mean that it's in any way lucrative. We find that

$$E = (1,000,000) \cdot \frac{8}{311,875,200} = \frac{8,000,000}{311,875,200} = \frac{2500}{97,461} \approx \$.0257,$$

or approximately $2\frac{1}{2}$¢.

If someone were to offer you 3¢ for the opportunity to take your spin on the machine—again ignoring all other payoffs—you would be wise to accept it, for they're offering you more than this chance is worth.

Example 4.2.5. Example 2.3.1 described the carnival game where the barker attempts to guess the month of your birth. Noting right away that your chance of winning is greater than your chance of losing, how does the carnival profit from this game?

The expectation of this game rests on two factors: the cost to play and the value of the prize. Let us denote these two values by C and V. If you win the game, you win a prize valued at V, but must subtract the cost C of playing from its value. If you lose, of course, your loss is $\$C$. The expectation from this game is then

$$E = (V - C) \cdot \left(\frac{7}{12}\right) + (-C) \cdot \left(\frac{5}{12}\right) = \frac{7V}{12} - C,$$

and as long as this value is negative, the carnival has an advantage. This happens when

$$V < \frac{12C}{7},$$

and so if, for example, the player is charged $3 for a chance at a prize, the prizes can be worth as much as $5.14 each and the carnival will still make money in the long run.

An important principle of expected value, one that will be very important in the analysis of gambling systems, is contained in the following theorem.

Theorem 4.2.1. *If* $X_1, X_2, \ldots, X_n$ *are random variables, then*

$$E(X_1 + X_2 + \cdots + X_n) = E(X_1) + E(X_2) + \cdots + E(X_n).$$

Another way to state this result is to say that *expectation is additive*.

Roulette

Example 4.2.6. Let X be the random variable measuring the outcome of a $1 bet on the number 17 at American roulette. The probability distribution for X is

x	-1	35
$P(X = x)$	$\frac{37}{38}$	$\frac{1}{38}$

and the expected value of X is

$$E(X) = (-1)\left(\frac{37}{38}\right) + (35)\left(\frac{1}{38}\right) = -\frac{2}{38} \approx -\$.0526 = -5.26\text{¢}.$$

When gambling, $E(X)$ represents a "typical" outcome. In a single spin of a roulette wheel, we will never lose exactly 5.26¢ on a $1 bet, but if we make a large number of bets on many successive spins, we will find that our average loss per spin will be very close to this value.

In a casino game, the expected value of a random variable has an important interpretation as the *house advantage* (abbreviated HA), or the percentage of each bet that a casino expects to win. For convenience, we will assume that most bets are for $1. This makes converting house advantages to percentages easier. For bets of more or less than $1, the expected value can be found by multiplying both sides of the equation by the amount wagered, and the HA can be found by dividing the calculated expectation by the size of the bet.

If the expected value is negative, then the game favors the casino; if it is positive, the game favors the gambler. Very few casino games—blackjack with card counting, video poker under certain payoff tables and with perfect strategy, live poker, and sports betting—ever have a favorable long-term edge for the player. It should be noted that these four games all have a skill component. No casino game where the results are completely due to chance has a positive player expectation under ordinary playing conditions—although see Example 5.2.12, where a casino error created a lucrative opportunity for players. Many of the house edges computed in this and the next chapter are collected in the appendix on page 245.

Example 4.2.7. If we consider the bet in Example 4.2.6 on a European roulette wheel, the only change that needs to be made is switching 37 for 38 in the denominators to reflect the absence of a 00 on a European wheel. The resulting expectation is

$$E = (-1)\left(\frac{36}{37}\right) + (35)\left(\frac{1}{37}\right) = -\frac{1}{37} \approx -\$.0270,$$

for an HA of 2.70%, about half the advantage of American roulette. While both games favor the casino, the player has a better chance of winning if the 00 is absent.

Example 4.2.8. More generally, for any bet on n numbers on an American roulette wheel except the basket bet $(n = 5)$, we have this probability distribution:

x	-1	$(36 - n)/n$
$P(X = x)$	$\frac{38-n}{38}$	$\frac{n}{38}$

which corresponds to an expectation of

$$E = \left(\frac{36 - n}{n}\right) \cdot \left(\frac{n}{38}\right) + (-1) \cdot \left(\frac{38 - n}{38}\right) = -\frac{2}{38}.$$

This is independent of n and shows that the casino has a constant 5.26% edge on every roulette bet other than the basket bet.

With this house advantage on every bet except the basket bet, where the HA is 7.89%, American roulette is one of the worst games in a casino for players. It is an excellent example of the gambler's rule of thumb expressed on the "Luck Be An Old Lady" episode of *Sex and the City* (2002):

> *The easier a bet is to understand, the greater the house edge.*

It's not difficult at all to understand the idea of "pick some numbers and see if they come up on a wheel." By contrast, the house advantage on a blackjack hand, for a player using perfect basic strategy, is approximately .5% (depending on the number of decks used and the exact house rules), and the HA on the pass line bet in craps is only 1.41% (see Example 4.2.10). The calculations leading to these values are somewhat more complicated than those above, as are the rules of the games.

Dice Games

Example 4.2.9. In the game of chuck-a-luck described in Example 4.1.7, a $1 bet pays $1 for each time the selected number appears on the dice, so a player can win $1, $2, or $3—or lose $1. A winning player's original bet is returned, so we can modify the probability distribution for X, the number of times the selected number appears, to a distribution for the new random variable Y that counts a player's winnings. If $X = 0$, then $Y = -1$; all other values for Y are identical to those for X.

y	-1	1	2	3
$P(Y = y)$	$\frac{125}{216}$	$\frac{75}{216}$	$\frac{15}{216}$	$\frac{1}{216}$

The expected value of a $1 bet on a single number in chuck-a-luck is thus

$$E(Y) = (-1)\left(\frac{125}{216}\right) + (1)\left(\frac{75}{216}\right) + (2)\left(\frac{15}{216}\right) + (3)\left(\frac{1}{216}\right)$$

$$= -\frac{17}{216} \approx -\$0.0787,$$

indicating a house advantage of about 7.87%.

This may be the reason for chuck-a-luck's disappearance from casinos—if the house advantage is too high, players will eventually avoid the game. Individual bets with a high house advantage are not a barrier to long-term player acceptance of a casino game—many bets on the craps table, for example, have edges higher than 7.87%. In chuck-a-luck, however, the only bets available are single-number bets, and so all bets have the same 7.87% house edge. Without a less unfavorable betting option, chuck-a-luck has little to offer a perceptive gambler.

Example 4.2.10. In craps, we saw in Example 3.2.6 that the probability of winning a bet on the pass line is $p = \dfrac{244}{495} \approx .492$. For a \$1 bet, the expected return is

$$E = (1)\left(\frac{244}{495}\right) + (-1)\left(\frac{251}{495}\right) = -\frac{7}{495} \approx -\$.0141 = -1.41\cent,$$

so the house advantage is a mere 1.41%. The corresponding expectation for the don't pass bet, incorporating the Bar 12 rule, is

$$E = (1)\left(\frac{949}{1980}\right) + (0)\left(\frac{1}{36}\right) + (-1)\left(\frac{244}{495}\right) = -\frac{3}{220} \approx -\$.0134 = -1.34$$

so the HA is 1.34%.

These results show that, from a purely mathematical standpoint, the house advantages on pass and don't pass bets are very low, and these bets stand among the best in the casino for the gambler. These low house advantages can be reduced further by the use of the *free odds* betting option, which we shall consider in Chapter 5.

Many of the other bets on a standard craps layout are simple one-roll bets on what the next roll of the dice will be. These bets have the advantage of being easy to understand, but as we might now expect, their house advantage is large.

Example 4.2.11. The "Any Seven" or "Big Red" bet is a one-roll wager that the next roll of the dice will show a 7. This bet pays off at 4 to 1 (or 5 for 1, as shown in Figure 3.1), and the probability of winning is just the probability of rolling a 7: $\frac{1}{6}$. The expected return on a \$1 Any Seven bet is

$$E = (4)\cdot\left(\frac{1}{6}\right) + (-1)\cdot\left(\frac{5}{6}\right) = -\frac{1}{6} \approx -\$.1667,$$

so the HA is 16.67%.

The alternate name "Big Red" is a pretty good description of where your bankroll is headed if you make this bet, and in light of the other available bets with less onerous HAs, this one should be avoided.

Example 4.2.12. In Example 3.2.8, we looked at the C&E bet, which combines the Any Craps and Eleven bets. Assuming a $2 bet split evenly between the two wagers, what is the HA of C&E?

The Any Craps bet pays off at 8 for 1 (or 7 to 1) while Eleven's payoff is 15 for 1 (14 to 1). Note that if one of the two bets in this combination wins, the other loses; this must be accounted for in the calculations.

If a 2, 3, or 12 is rolled, the net win is $7 − $1, or $6. This event has probability 4/36. If the dice show 11, the net payoff is $14 − 1 = $13, with probability 2/36. The probability distribution for the net winnings X on the C&E bet is therefore

x	6	13	−2
$P(X = x)$	$\frac{4}{36}$	$\frac{2}{36}$	$\frac{30}{36}$

and the expectation is

$$E = (6) \cdot \left(\frac{4}{36}\right) + (13) \cdot \left(\frac{2}{36}\right) + (-2) \cdot \left(\frac{30}{36}\right) = -\frac{10}{36} \approx -\$.2778.$$

Dividing this expectation by the $2 wagered gives an HA of approximately 13.89%—better than "Any Seven," but still too high for the serious craps player.

Example 4.2.13. Another one-roll bet popular on craps tables is the *field* bet. By placing a bet in the Field section of the layout, the bettor is wagering that the next roll will be 2, 3, 4, 9, 10, 11, or 12. Many casinos will pay double on this wager if the roll is a 2 or 12; the bet pays even money otherwise.

On the surface, this looks like a good bet—a field bet covers 7 of the 11 possible sums and pays double on two of them. Of course, as we know, this apparent advantage is based on an incorrect interpretation of the sample space (see Example 2.2.2). On a $1 bet:

- There are two ways to win $2: by rolling 1-1 or 6-6.

- There are 14 ways to win $1: when the roll is a 3 (two ways), 4 (three ways), 9 (four ways), 10 (three ways), or 11 (two ways).

- The remaining 20 rolls—totals of 5, 6, 7, or 8—result in a $1 loss.

With the facts before us, we can derive the probability distribution for the net amount won on a $1 field bet:

x	−1	1	2
$P(X = x)$	$\frac{20}{36}$	$\frac{14}{36}$	$\frac{2}{36}$

and compute the expectation:

$$E = (2) \cdot \left(\frac{2}{36}\right) + (1) \cdot \left(\frac{14}{36}\right) + (-1) \cdot \left(\frac{20}{36}\right) = -\frac{1}{18} \approx -\$.0556.$$

The HA on this bet is therefore 5.56%—exactly one-third the HA of "Any Seven."

A second class of craps bets can extend over several rolls before being resolved. The *hardway* bets are one example. A hardway bet can be made at any time, whether or not a point has been established. A player making a hardway bet on an even number (4, 6, 8, or 10) bets that that number will be rolled "the hard way," as doubles, before it is rolled another way or a 7 is rolled.

A hardway bet can be won if any of the following sequences of events happens:

- The number is rolled as doubles on the first roll after the bet is made.

- Neither the number nor a 7 is rolled on the first roll, and the number is rolled as doubles on the second roll.

- Neither the number nor a 7 is rolled on the first two rolls, and t number is rolled as doubles on the third roll.

- Neither the number nor a 7 is rolled on the first three rolls, and the number is rolled as doubles on the fourth roll.

- And so on, through an infinite number of possibilities. As the number of rolls increases, of course, the probability of needing that many rolls to resolve the bet decreases.

Since the successive rolls are independent, we could use the Multiplication Rule to compute the probability of each of these sequences and the mathematical theory of infinite series, which is beyond the scope of this book, to add them all up. As we saw with the pass and don't pass bets in Chapter 3, an easier approach to computing probabilities for hardway bets may be used. It focuses only on the rolls that will resolve the bet.

Example 4.2.14. Consider the case of a hardway bet on 6, which pays off at 9 to 1. We can ignore all possible rolls except 6s and 7s—there are 11 of these to be considered. Only one of them, the 3-3, results in this bet winning, so the probability of winning the bet is $\frac{1}{11}$. The expected value of a $1 bet is then

$$E = (9) \cdot \left(\frac{1}{11}\right) + (-1) \cdot \left(\frac{10}{11}\right) = -\frac{1}{11} \approx -\$.0909,$$

for a 9.09% house advantage. This bet may be better than "Any Seven," but there are still far better bets available. Even the field bet is a better bet for the player.

Since 6 and 8 are equally likely, the HA of a hardway bet on 8 is a 9.09%. Hardway bets on 4 or 10 pay off at 7 to 1, a lower payoff since there are fewer "easy" ways to roll these numbers. Confining our attention to the rolls that resolve the bet, we find that the probability of winning this bet is $\frac{1}{9}$, from which it follows that the house edge is 11.11%.

Example 4.2.15. In Die Rich (Example 3.2.21), we had a table of outcomes and payoffs that results in the following probability distribution:

x	1	2	−1
$P(X = x)$	$\frac{13}{54}$	$\frac{13}{81}$	$\frac{97}{162}$

The expectation for a \$1 bet, taking the varying payoffs into account, is

$$E = (1) \cdot \left(\frac{13}{54}\right) + (2) \cdot \left(\frac{13}{81}\right) + (-1) \cdot \left(\frac{97}{162}\right) = -\frac{1}{27} \approx -\$.0370,$$

which yields a house advantage of about 3.70%. This makes Die Rich a worse bet than a Pass or Don't Pass bet at standard craps, but ranks it ahead of most other craps bets for the player.

Bingo, Keno, and Lotteries

Example 4.2.16. In an actual Cover All bingo game (Example 3.2.15), it is highly unlikely that a game will last until all 75 numbers are drawn. This would require that every player in the game have a number common to all of their cards, and that that number be the last one drawn—which would result in a massive tie as everyone shouted "Bingo!" at once. We can compute the average number of draws required for a given card to be covered by evaluating the sum

$$\sum_{k=24}^{75} k \cdot P(\text{Win in } k \text{ draws}) = \sum_{k=24}^{75} k \cdot \frac{24 \cdot 51! \cdot (k-1)!}{(k-24)! \cdot 75!} = 72.96.$$

If 200 cards are in play, then repeated simulation suggests that it will take about 62 numbers before a winner is determined. This number slowly decreases with the number of cards in a game; for 1000 cards, the average is 58.85 [71].

If a game does not refund the original bet to winning players, we must take this into account when computing expected returns. This is the case, for example, when playing keno or buying a lottery ticket. The easier way to do this is to reduce the advertised payoffs by the cost of the wager, and the next few examples demonstrate this approach.

Example 4.2.17. The Michigan State Lottery's "Club Keno" game, played in bars across the state, is an electronic version of keno. The 1 Spot Game asks the player to pick one number in the range from 1-80. The state then draws 20 numbers in that range, and if the player's number is among the 20, the payoff is $2. The probability of winning this bet is the probability of one's number appearing as one of the 20 drawn. We compute this by looking at the other 19 numbers, which may be chosen from 79 possibilities, so:

$$P(\text{Win}) = \frac{{}_1C_1 \cdot {}_{79}C_{19}}{{}_{80}C_{20}} = \frac{1 \cdot \frac{79!}{60! \cdot 19!}}{\frac{80!}{60! \cdot 20!}} = \frac{79! \cdot 60! \cdot 20!}{80! \cdot 60! \cdot 19!} = \frac{1}{80} \cdot 1 \cdot \frac{20}{1} = \frac{20}{80} =$$

The probability of losing is therefore $\frac{3}{4}$. If tickets cost $1, which is paid when the ticket is purchased, then the appropriate probability distribution for the player's winnings X is

x	−1	1
$P(X = x)$	$\frac{3}{4}$	$\frac{1}{4}$

where each of the outcomes—$0 or $2—has been reduced by the cost of the ticket. The corresponding expected value is

$$E = (-1)\left(\frac{3}{4}\right) + (1)\left(\frac{1}{4}\right) = -\$.50 = -50\text{¢}$$

—which makes this bet equivalent to asking someone for change for a dollar and receiving two quarters in return. And being happy with the exchange.

Example 4.2.18. For the keno game described in Example 2.4.20, we saw that the probability of winning was .0616. Of course, part of the lure of keno is the potential to win a large payoff for a small bet, as when matching seven out of seven numbers (an admittedly rare occurrence) pays off $8000 for every dollar bet. Do the higher payoffs for the rarer events result in a reasonable expectation?

Let Y be the net amount won on a $1 bet, taking the cost of the ticket into account. The probability distribution for Y is

y	−1	0	19	299	7999
$P(Y = y)$	.9384	.0522	.0086	.00073	2.44×10^{-5}

The corresponding expectation is

$$E = (-1)(.9384) + (19)(.0086) + (299)(.00073) + (7999)(2.44 \times 10^{-5})$$
$$\approx -\$.363,$$

so the house advantage is about 36%, and while there is no well-defined meaning of "reasonable bet," it is safe to say that this bet does not qualify.

Another option when incorporating the cost of the wager into an expected value calculation is to work with the payoffs as advertised and then, at the end, to subtract the price of the ticket from the sum. The following two examples illustrate this method.

Example 4.2.19. A *50/50 drawing* is a common version of a lottery, often used as a charitable fundraiser, where 50% of the money raised by selling tickets is returned as a prize to the holder of the winning ticket. If the tickets are sold for $1 each, and n tickets are sold, then the prize will be $\$\frac{n}{2}$ and the expected win when buying a single ticket is

$$E = \left(\frac{n}{2}\right) \cdot \frac{1}{n} = \$.50.$$

From this amount, which is independent of the number of tickets sold, we subtract the cost of the ticket to arrive at a final expectation of $-\$.50 = -50\cent$.

An expected value of half the ticket price is what we would expect, given the prize structure of a 50/50 drawing. This extends to the case where you purchase k tickets instead of just 1; while your probability of winning rises to $\frac{k}{n}$, you must subtract $\$k$ at the end rather than just $\$1$, and so your expectation is $-\$k/2$, or half of the money you paid out at the start.

Example 4.2.20. In the Meskwaki Casino's Super 20 Special (Example 4.1.8), the expected *number* of matches is

$$\sum_{x=0}^{20} x \cdot P(X = x) = \sum_{x=0}^{20} x \cdot \frac{{}_{20}C_x \cdot {}_{60}C_{20-x}}{{}_{80}C_{20}} = 5,$$

and so it is no surprise that the casino has tagged this event, and the two nearest to it, for a player loss. The expected value of a $5 bet can be computed by using the payoff table for this game:

Matches	0	1	2	3	4	5	6
Payoff	500	10	5	5	0	0	0
Matches	7	8	9	10	11	12	13
Payoff	5	5	25	50	200	1000	4000
Matches	14	15	16	17	18	19	20
Payoff	7500	18,000	35,000	45,000	55,000	65,000	100,000

Denoting this payoff function by $A(x)$, where x is the number of matches, and subtracting $5 at the end to incorporate the cost of the ticket into the expectation, we find that

$$E = \left[\sum_{x=0}^{20} A(x) \cdot P(X = x)\right] - 5 \approx -\$1.73,$$

which, upon dividing by the cost of the ticket, gives an HA of 34.6%.

Example 4.2.21. The Michigan State Lottery also offers a "Daily 3" game, for which numbers are drawn twice a day, seven days a week. The simplest bet is a "straight" bet, where the gambler picks a three-digit number and wagers $1. If the player's number matches the three-digit number drawn by the state, the payoff is 500 for 1. Notice that the payoff is "500 *for* 1" rather than "500 *to* 1" and so the profit for a winning player is only $499. Since there are 1000 possible three-digit numbers, from 000 through 999, the probability of winning this lottery is $\frac{1}{1000} = .001$. The expected value of a $1 bet is

$$E = (499) \cdot \frac{1}{1000} + (-1) \cdot \frac{999}{1000} = -\frac{500}{1000} = -\$.50$$

—and so the state takes half of every dollar wagered on a Daily 3 ticket.

Example 4.2.22. Another Daily 3 option is the box bet. As in horse racing (Example 2.4.11), when a player "boxes" a three-digit number, the ticket wins if the three digits comprising that number come up, in any order, in the state's chosen number. For example, if you make a box bet on the number 678, you win if the state's number is 678, 687, 768, 786, 867, or 876. The payoff structure depends on whether the player's number contains two or three different digits; we shall consider the case where the three digits are all distinct. This leads to what is called a "six-way boxed bet," since there are $_3P_3 = 6$ different permutations of a three-digit number consisting of three different digits. This bet pays off at 83 for 1 if any of the six permutations is drawn, and so the expected value is

$$E = (82) \cdot \frac{6}{1000} + (-1) \cdot \frac{994}{1000} = -\frac{502}{1000} = -\$.502,$$

slightly less than the expectation for a $1 straight bet.

While the lottery bets considered in Examples 4.2.17, 4.2.19, 4.2.21, and 4.2.22 are exceptionally bad bets from the gambler's perspective, one fact that must be addressed here is that state lotteries—like most 50/50 drawings—are intended in part as a fundraising mechanism. Since its inception in 1972, the Michigan State Lottery has contributed over $16 billion to the state's public schools, and that number is far greater than it would be if players were getting a 95% return on their lottery dollars [47]. A 50% house advantage would be untenable in a casino, but is readily accepted in a state lottery when it is understood that a portion of the proceeds is supporting education.

Daily state lottery drawings can trace their origins to an underground game called the *Numbers Game*, an illegal form of gambling formerly popular in large cities [60]. The Numbers Game is a very simple proposition: players bet on a number of their choice and win their bet if the number appearing in some legitimate and presumably unbiased source—for example, the U.S. Treasury balance as published in a newspaper or the payoff price at an indicated racetrack—matches theirs. In its heyday, the Numbers Game attracted millions of dollars in bets and engaged a significant fraction of the adult population in some areas.

Example 4.2.23. In New York City, one popular Numbers game was the "You Pick 'Em Treasury Ticket" [60, p. 168–9]. Players chose a five-digit number and were paid off in accordance with the last five digits of the published U.S. Treasury balance each day. If a player matched all five digits, the payoff was 300 for 1; matching only the last four paid 30 for 1 and a match of the last three paid 3 for 1.

There are 100,000 different five-digit numbers, so the odds against the player are immense. There is 1 number that is an exact match, 9 numbers that match only the last four digits, and 99 that match only the last three. The expected value of a $1 Treasury ticket is then

$$E = (299) \cdot \frac{1}{100,000} + (29) \cdot \frac{9}{100,000}) + (2) \cdot \frac{99}{100,000} + (-1) \cdot \frac{99,891}{100,000}$$
$$= -\frac{99,133}{100,000},$$

or −99.133¢—a 99.133% house advantage.

As awful as the player's disadvantage may be in a state lottery, it's far better than it was in the days of the Numbers Game. Despite the long odds, the Numbers Game was quite popular. In Detroit, an employee at the *Free Press* once clipped the two lines containing the day's Treasury balance from the business pages so that the stock market tables would all fit. When the newspaper hit the streets the next morning, the paper's telephone lines were swamped with calls wanting to know where the numbers were [16].

The Big Six Wheel

Example 4.2.24. Is the Big Six wheel (Example 2.3.2) lucrative? Only for the casino, as we might expect given the simplicity of the game. The standard Las Vegas rules for the Big Six wheel state that each currency amount pays its face value per $1 bet, and the two logos each pay off at 40 to 1 [67]. Armed with these payouts, we can see how much of an edge the casino has. For a bet on the $1 spot:

$$E = (1) \cdot \left(\frac{24}{54}\right) + (-1) \cdot \left(\frac{30}{54}\right) = -\frac{6}{54} = -\frac{1}{9} \approx -\$.1111.$$

This 11.11% house advantage is the best bet, for the player, on the Big Six wheel. Other bets have HAs ranging from 16.67% to 24.07% (see Exercise 4.11).

In New Jersey, state regulations require that the two logos pay off at 45 to 1, rather than 40 to 1. While the HA on these bets drops from 24.07%

to 11.11%, a 2011 study [63] showed that one casino's hold (the percentage of wagered money that is kept by the casino) at the Big Six wheel averaged 42.25%, well in excess of even the highest house advantage. This increased casino hold arises due to players re-betting their winnings and thus exposing more money to the already-high house advantage.

This study confirmed that Big Six is not a player-friendly game in practice, as well as in theory.

Baccarat

Baccarat (bah-kuh-RAH, from the Italian *baccara*, "zero") is a card game similar to blackjack that is often played in the high-limit area of casinos and is favored by many high rollers, among whom it is not unusual to risk tens of thousands of dollars on a single hand. Outside high-limit areas, it is seen in a scaled-down form known as *mini-baccarat* or "mini-bac" for short. In either case, the rules are simple: gamblers bet on which of two hands—called the *Player* and the *Banker*—will be closer to 9. These bets are paid at 1 to 1. A third option is to bet that the two hands will tie, which pays off at 8 to 1. If the hand is a tie, bets on Banker and on Player are also ties, and no money is won or lost on them.

Two cards, from a shoe (a box that holds multiple decks of playing cards) containing six to eight decks, are initially dealt to each hand. The Riviera Casino in Las Vegas once used a 16-deck shoe in an effort to encourage players to gamble longer—many players leave almost re-flexively at the conclusion of a shoe while the cards are being shuffled. The Union Plaza Casino in downtown Las Vegas (now the Plaza) experimented briefly with what was billed as "the world's largest baccarat shoe" (see Figure 4.1), which held 144 decks, or 7488 cards, but this novelty was short-lived [81, p. 144]. With a single deck of cards measuring about 1.6 centimeters thick, this monster shoe measured about 7.5 feet long, held nearly 30 pounds of cards, and would certainly have been a challenge for the dealers to use.

FIGURE 4.1: A $3 Union Plaza baccarat chip advertising their 144-deck shoe

Hand values are computed by counting each card at its face value, with aces counting 1 and face cards 0. If the sum of the cards exceeds 9, then the tens digit is dropped. For example, a hand consisting of a 7 and a 5 has value 2. Unlike in blackjack, a baccarat hand does not "bust" if it exceeds the highest possible hand value of 9.

If either hand has a value of 8 or 9, the hand is called a *natural*, and no further cards are drawn. If there are no naturals, an intricate set of rules dictates when either hand may receive a third card. The Player hand, which always goes first, draws a third card if its value is 5 or less. If the Player stands with 6 or 7, the Banker hand draws a third card if its value is 5 or less. If the Player draws a third card, the Banker hand's action is determined by its value and the value of the Player's third card. Table 4.3 (from [51]) shows the standard set of rules.

TABLE 4.3: Baccarat rules for the Banker hand

Banker's hand	Banker draws if Player's 3rd card is
0–2	Any
3	Not an 8
4	2–7
5	4–7
6	6 or 7
7	None

Example 4.2.25. Suppose that the cards dealt to the Player hand are K and $7\diamondsuit$, and that the Banker hand receives $7\spadesuit$ and $6\diamondsuit$. The Player hand totals 7, and so does not draw a third card. Since the Banker's total is 3, which is less than 5, a third card is dealt to the Banker hand. If this card is the Q the hand remains a 3, and so the Player wins, 7 to 3.

Example 4.2.26. In the next hand, suppose the Player's hand is $9\heartsuit$ and 2 for a total of 1, and the banker is dealt $K\spadesuit$ and $4\heartsuit$, totaling 4. The Player hand draws a third card, the $6\spadesuit$, for a total of 7. Since the Player's third card was a 6, a Banker hand of 4 is required to draw a third card. If that card is the $6\clubsuit$, the Banker hand is now 0, and Player wins, 7 to 0.

These rules include some cases where the Banker hand is required to draw another card even though it's beating the Player hand already. For example, suppose that the Player hand is A-3, totaling 4. The Player takes a third card and draws a 7, bringing that hand's total to 1. If the Banker's hand is 9-6, for a total of 5, the Banker leads but must nonetheless take another card since the Player's third card was a 7. It is also possible for the Banker hand to be denied a third card even if it's tied with the Player hand, as in the case where the Player has 2-3 and draws an 8 for a total of 3, and the Banker holds 3-K. Because the Banker has 3 and the Player's third card was an 8, no further card will be drawn to the Banker hand, and the round ends in a tie.

How did game designers settle on these rules? Consider the rule cited above for drawing a third card to a Banker hand of 3. If the Player hand has drawn an 8 as its third card, it follows that the hand's value is 8, 9, 0, 1, 2, or 3, according as the initial two-card hand was 0, 1, 2, 3, 4, or 5. We may assume that these six totals occur equally often in the long run. If the Banker hand is 3, it wins half the time and ties one-sixth of the time, and the desire for a balanced game makes this a good place to stop drawing cards.

Since the rules of baccarat are fixed, it is entirely a game of chance. Contrary to the image of baccarat as presented in popular culture, such as when James Bond plays the game in six movies, skill plays no part in the play of the game. The replacement of baccarat by Texas Hold'em in the 2006 version of *Casino Royale* represents something of a betrayal of the Bond legend even as it tapped into a wave of poker popularity.

In light of these complicated rules and the maxim that "the easier a bet is to understand, the higher the house advantage," we might expect baccarat to have a fairly low HA, and indeed this is so. In an eight-deck game, the probability of winning a resolved bet (that is, when the hand is not a tie) on Player is .4932 and the probability of winning a resolved bet on Banker is .5068 [51]. As with the Don't Pass bet in craps, it is not in the casino's best interests to offer a bet with a positive expected value, and so it is standard practice for casinos to charge a 5% commission on winning Banker bets. In effect, this means that if you win a bet on Player, you are paid at 1 to 1 odds, but if you bet on Banker and win, you receive only 95% of the amount wagered (19 to 20 odds).

As a convenience to everyone involved, players and dealers alike, some casinos use special $20 chips at their baccarat tables; this is a denomination not typically used elsewhere in the casino. These chips make the calculation of commissions easy: $1 for every $20 chip in play on a winning Banker wager. In practice, to avoid complicated change-making transactions, all winning Banker bets are paid at even money, and commissions owed the casino are simply counted during the play of a shoe. When a player wishes to leave or the round has ended and the cards are being shuffled or replaced, the casino collects the accumulated commissions owed.

Example 4.2.27. As is also the case in blackjack, the probabilities of winnin baccarat wagers depend on the number of decks of cards being used. If eight decks are used, these rules lead to the following probability distributions:

- Let X be the return on a $1 Player bet. The probability distribution for X is

x	−1	1	0
$P(X = x)$	.4584	.4461	.0955

and the corresponding expectation is

$$E(X) = (-1)(.4584) + (1)(.4461) + (0)(.0955) = -\$.0123,$$

so the house advantage is 1.23%.

- If Y is the return on a $1 Banker bet, the distribution for Y is

y	-1	.95	0
$P(Y = y)$	.4461	.4584	.0955

and the expected value is

$$E(Y) = (-1)(.4461) + (.95)(.4584) + (0)(.0955) = -\$.0106,$$

for an HA of 1.06%.

- Finally, if Z is the return from a \$1 Tie bet, the payoff is 8 to 1, and the distribution is

z	-1	8
$P(Z = z)$	.9045	.0955

The resulting expected value is then

$$E(Z) = (-1)(.9045) + (8)(.0955) = -\$.1405,$$

for an HA of 14.05%—this is clearly a bet to be avoided.

As a promotional tool intended to attract bettors to the baccarat tables, casinos have been known to experiment with lowering the commission charged on winning Banker bets, which, of course, reduces the HA. If the commission is reduced to 4%, the house advantage drops to .60%, and at a 3% commission, the HA is a nearly invisible .14%—the game is essentially even at that level, as a \$100 bet loses, on the average, only 14¢.

In considering the mathematical aspects of baccarat, we are immediately drawn to one difference between it and other games we have considered: successive hands are not independent. As is also the case in blackjack, a card dealt in one hand is a card that cannot be dealt in the next hand, and so the probabilities of different hand totals change as the dealer moves through the shoe. That being the case, a mathematical analysis of baccarat requires somewhat more involved calculations that account for the changing deck composition. The following extended example will illustrate this.

Example 4.2.28. A variation of the baccarat stand or draw rules gives the gambler making the largest Player wager the option of deciding whether a Player hand of 5 draws a third card or stands. This less common version, sometimes called *chemin de fer*, appears to have a skill component, but does it really, or is the casino offering the player a meaningless choice?

That question can be answered with some simple reasoning, and aside from accounting for the lack of independence between hands, requires very little mathematics. Assuming that the Banker hand is not a natural, a Player hand of 5 wins against Banker hands of 0 to 4, ties a Banker 5, and loses only against a 6 or 7. The player making this decision has full knowledge of the Banker hand and a large bet on Player, and so certainly should choose to draw when losing to a 6 or 7—as the standard baccarat rules provide. In chemin de fer, the Banker hand will not draw to a 6 or 7 if the Player hand does not take a third card, and so will win if the player chooses not to draw.

Against a Banker hand of 0 to 4, the player in control of the decision must weigh the possibility of improvement against the fact that the hand is winning as dealt. If the player chooses never to draw to a 5, the Banker hand will draw another card, and so has a chance to convert a losing or tying hand into a winning hand. There is no instance where a winning Banker hand is converted to a nonwinning hand when the player stands on 5—at worst, a losing Banker hand remains a loser; it might improve and tie or win. It follows that standing on a 5 in these circumstances does not improve the Player hand's chance of winning.

If the hands are tied and the Player hand stands on 5, then the Banker hand will draw a third card. A decision by the player not to draw to the Player hand is a gamble that the next card out of the shoe will not improve a 5. While the exact probability of improvement depends on the cards that have been dealt, including the exact cards that constitute the two dealt 5s, we can get a sufficiently accurate estimate of the probability by assuming a full shoe. If the player chooses to draw a third card to a 5:

- The hand is improved if the third card dealt to it is an ace through 4, and the probability of improvement is therefore $20/52 = 4/13$. If that third card is not a 4, then Table 4.3 indicates that the Banker hand does not draw again, and so the Player wins with probability $3/13$.

- The hand stays the same if a face card or 10 is drawn, an event wh also has probability $4/13$. The Banker hand does not draw under these circumstances, and so the hand remains tied.

- The hand loses value—dropping below 5—on any other card: a 5 through 9. This has probability $5/13$. If the third card is an 8 or 9, then the Banker hand does not draw a third card, and so wins automatically. On a 5, 6, or 7 (final Player hand of 0, 1, or 2), the Banker must draw again.

We have the following probabilities:

Event	Probability
Player wins	3/13
Banker wins	2/13
Hand is a tie	4/13
Banker draws	4/13

We shall break down the four subcases, each with probability $1/13$, that make up the "Banker draws" outcome. If the Player's third card is a 4, then the

Banker hand is drawing against a Player 9, and so cannot win. The Banker ties only if the third card drawn to that hand is also a 4. The probability of drawing a 4 depends on how many 4s are included among the five cards already played; we know that that number is at least 1. An upper bound for the probability of a tie, assuming an eight-deck shoe, is $31/411 \approx .0754$, since there are no more than 31 4s left among the 411 undealt cards. The maximum number of 4s removed from the deck is 3, in the case where each hand has been dealt a 4 and an ace initially, and so the probability of a tie is no less than $29/411 \approx .0706$. ($.0706$–$.0754$ will also be the range of the probability of a tie in the other situations where the Banker must draw, because in each case we are simply computing the probability that the Banker's third card matches the Player's third card.) The probability that the Player hand wins is then between $380/411 \approx .9246$ and $382/411 \approx .9294$.

In the other cases where the Banker hand must draw a third card, the Banker can win or tie, and can lose if the Player hand is not 0. The ranges of the various probabilities are collected in the following table:

| Player | | Outcome | | |
Third card	Total	P(Banker win)	P(Player win)	P(Tie)
4	9	0	.9246–.9294	.0706–.0754
5	0	.9246–.9294	0	.0706–.0754
6	1	.8467–.8540	.0730–.0779	.0706–.0754
7	2	.7689–.7762	.1509–.1557	.0706–.0754

The "Player draws 4" and "Player draws 5" cases are mirror images of one another; in the other two cases, the Banker has a large edge over the Player. How does this edge compare to the edge of $1/13$ that the Player has in all other cases?

We shall consider the case where neither the Banker nor the Player hands contains a card of the same rank as the Player's third hand. Under this assumption, the probability of a Banker win is

$$p_B = \left(\frac{1}{13}\right) \cdot (.9294 + .8540 + .7762) = .1969,$$

the probability of a Player win is

$$p_P = \left(\frac{1}{13}\right) \cdot (.9294 + .0779 + .1557) = .0895,$$

and the probability of a tie is

$$p_T = \left(\frac{4}{13}\right) \cdot .0706 = .0217.$$

Combining this information with the probabilities in the first three lines of the table above gives the following probabilities of the three possible outcomes when the player chooses to draw to a 5 against a Banker 5.

Event	Probability
Player wins	$3/13 + .0895 \approx .320$
Banker wins	$2/13 + .1969 \approx .351$
Hand is a tie	$4/13 + .0217 \approx .329$

As a result, the expectation of a \$1 bet in this situation is

$$E = (1) \cdot .320 + (-1) \cdot .351 + (0) \cdot .329 = -\$.031.$$

This value must be compared to the expected value when the player stands on 5. There are three possible outcomes.

- *The Player wins despite the Banker's draw.* This happens when the third card dealt to the Banker hand is a 5 through 9. The number of such cards left in the shoe is between 156 (if, for example, both Player and Banker hands are an 8 and a 7) and 160 (as is the case when both hands con of an ace and a 4, or a 2 and a 3). Since only four cards have been dealt, 412 remain in the shoe, and so

$$\frac{156}{412} \leq P(\text{Player win}) \leq \frac{160}{412}.$$

- *The hand remains tied.* The Banker hand remains a 5 if its third card is a 10 or a face card. There are between 126 and 128 10-count cards left in the shoe, as each of the Player and Banker hands can have at most 1 such card. It follows that

$$\frac{126}{412} \leq P(\text{tie}) \leq \frac{128}{412}.$$

- *The Banker hand draws a third card and wins.* The Banker's third card must be in the range ace–4. There are 124–128 such cards remaining, because the two hands dealt may contain, between them, 0, 2, or 4 of these low cards, and thus

$$\frac{124}{412} \leq P(\text{Banker win}) \leq \frac{128}{412}.$$

We can see immediately that at the start of a shoe, the probability that the Player wins when standing on 5 against a Banker 5 exceeds the probability of a Banker win. The Player's expectation is therefore positive when standing on 5 at the start of a shoe, and so standing has a higher expectation than drawing against a 5. Absent any useful information about the composition of the remaining cards, we have the following strategy for the lead bettor when the Player hand is dealt a 5:

Banker hand	Action
0–4	Draw
5	Stand
6–7	Draw

Since an optimal strategy exists, any deviation from this strategy increases the casino's advantage, and this may be a reason why a casino would offer this choice to a player—remember that the choice is given to the bettor making the largest Player bet, and so a choice that decreases the Player hand's chance of winning increases the likelihood that the casino will collect that largest wager.

Video Poker

While Texas Hold'em is the version of poker that has come to dominate both live play in casinos and televised play in poker tournaments, *video poker* is primarily a game of five-card draw poker simulated by a computer. As with slot machines, the player is playing against a fixed paytable rather than another human or electronic opponent. The goal is simply to end up with the best possible five-card poker hand: five cards are dealt, and the player may discard any or all of them and receive replacements.

There are two places in playing video poker where skill plays a part. The first, which begins before any money is put down, is selecting the best machine—in terms of the paytable offered—to play. The second, of course, lies in knowing what to discard and which hands to pursue in the draw. We begin with the first. The expected return for a video poker machine is typically expressed as its *payback percentage*, which with the right paytable and perfect play on the part of the player, can very occasionally exceed 100%—that is, the machine is expected to return more money than it takes in, in the long run. For example, Table 4.4 is a video poker paytable where all of the payoffs are "for 1," and include the amount wagered, so the 1 for 1 payoff for a pair of jacks or better is simply a break-even payoff [35, p. 121]. Most video poker machines accept wagers of one to five times the minimum bet; the amount wagered is typically described in the paytable as "n coins" even if the minimum amount does not correspond to a coin, as is the case, for example, with a $5 video poker machine.

TABLE 4.4: Jacks or Better video poker paytable with 99.8% return

Poker Hand	Payoff: 1 coin	Payoff: 5 coins
Royal flush	250	4700
Straight flush	50	250
Four of a kind	25	125
Full house	9	45
Flush	6	30
Straight	4	20
Three of a kind	3	15
Two pair	2	10
Pair of jacks or higher	1	5

It is customary to describe video poker paytables by the payoff for a one-coin bet on a full house and a flush; thus this would be a *9/6* game. A 9/6 machine is better than a 9/5 machine, which is in turn better than an 8/5 machine.

One thing to notice immediately is that, when betting five coins, the payoff on every hand but one is simply five times the payoff with a single coin at risk. The exception—and it's a big one—is the royal flush. This is important, since approximately 2% of the payback percentage on most video poker machines comes from royal flushes betting the maximum, or playing "max coins" for short [95]. By not ruthlessly pursuing royal flushes whenever the strategy tables advocate doing so, the player is giving up on a meaningful part of the payoff: reducing a 98% machine to a 96% one, for example. This means that not only should a video poker player bet max coins, but also that the drawing strategy should be tilted toward pursuing royal flushes, which have a probability of approximately 1 in 40,000 on a video poker machine. Contrast this with the probability of 1 in 649,740 of receiving a *dealt* royal flush, and you can see the value of being allowed to discard cards and replace them in the draw.

The payoff percentage for this paytable is 99.8%, assuming perfect play. What is perfect play? It depends on the machine and on the paytable; part of intelligent video poker play is learning the strategy appropriate to your machine. The ideal strategy for a given machine and paytable has been determined through computer simulation of millions of video poker hands. For the paytable above, the strategy in Table 4.5 leads to that 99.8% payback percentage. The corresponding 9/5 machine's payback percentage is 98.8%, and an 8/5 machine that is otherwise the same pays back at 97.6% [35, p. 121–122].

Table 4.5 should be read from the top down until you reach a combination contained in the hand you hold, and then you should discard appropriately and draw the number of cards indicated [35, p. 207].

The following terms are used in the table:

- An *outside straight* is four cards comprising a straight that is open at both ends, such as 6789. Drawing either a 5 or a 10 will complete this straight.

- An *inside straight* is four cards comprising a straight, but with a "hole" in the middle, such as 4568 or 5689. Only a drawn 7 will complete any of these straights, so an inside straight has less potential for improvement, and thus is less valuable, than an outside straight.

- *High* cards, for the purpose of jacks or better video poker, are the jack, queen, king, and ace—those cards that, when paired, pay off.

At a number of places in the table (line 4 and especially line 12), you can see that the pursuit of a royal flush is prioritized over lesser payoffs. Occasionally, perfect strategy may call on the gambler to break up a winning combination in

TABLE 4.5: Video poker strategy for standard 9/6 Jacks or Better

	When dealt	Draw
1:	Royal flush	0
2:	Straight flush	0
3:	4 of a kind	0
4:	4-card royal flush	1
5:	Full house	0
6:	Flush	0
7:	3 of a kind	2
8:	Straight	0
9:	2 pairs	1
10:	4-card straight flush	1
11:	High pair: J, Q, K, or A	3
12:	3-card royal flush	2
13:	4-card flush	1
14:	Low pair, 2–10	3
15:	4-card outside straight	1
16:	3-card straight flush	2
17:	Suited JQ, JK, or QK	3
18:	4-card inside straight (3–4 high cards)	1
19:	Suited JA, QA, or KA	3
20:	Nonsuited JQK	2
21:	Suited J10, Q10, or K10	3
22:	1 or 2 high cards	3-4
23:	5 mixed low cards (a *razgu*)	5

order to draw cards toward completing a royal flush. Unlikely as successfully completing a royal by drawing one or two cards may be, the high payoff at max coins leads to an expected value in excess of the hand being broken up.

Example 4.2.29. On a game played with Table 4.4 as paytable, if you are dealt

$$4\clubsuit\ 10\clubsuit\ J\clubsuit\ Q\clubsuit\ A\clubsuit,$$

then perfect video poker strategy dictates that you break up the flush by discarding the $4\clubsuit$ and draw one card, hoping to pull the $K\clubsuit$ needed to complete the royal flush. (In Table 4.5, line 4, 4-card royal flush, appears before line 6, Flush.) The greatly enhanced payoff for a royal flush with max coins wagered makes this the better play, as we can see by computing expected values. If you hold the $4\clubsuit$ and cash in the flush, your expected profit is \$25 with a max coins bet, assuming a \$1 coin. If you discard the 4 and go for the royal flush, the following outcomes are possible:

Result	Net Payoff	Probability
Royal flush	4695	1/47
Flush (nonroyal)	25	7/47
Straight	15	3/47
Pair (J, Q, or A)	0	9/47
Nonpaying hand	−5	27/47

The expected value of the hand if the 4♣ is discarded is

$$E = (4695) \cdot \left(\frac{1}{47}\right) + (25) \cdot \left(\frac{7}{47}\right) + (15) \cdot \left(\frac{3}{47}\right) + (0) \cdot \left(\frac{9}{47}\right) + (-5) \cdot \left(\frac{27}{47}\right)$$

$$= \frac{4780}{47} \approx \$101.70.$$

It follows that, even though the probability of winding up with nothing after the draw is over 50%, your expected return is over four times greater by giving up on the sure $25 payoff for the chance of hitting a royal flush.

Some video poker machines modify this paytable by paying out 4000 coins for a 5-coin bet with a royal flush. The payback percentage on such a machine drops to 99.5%. Such a machine has an expectation of $86.81 for this hand, which is less than $101.70 but still better than the $25 expectation of holding the dealt flush.

Example 4.2.30. What should you do if you are dealt the following hand?

$$5\diamondsuit \ J\clubsuit \ 4\spadesuit \ Q\spadesuit \ 3\diamondsuit$$

Reading down the chart, the first column entry that corresponds to your holdings is next to last: "1 or 2 high cards," so you should hold the jack and queen and draw three new cards.

Note that a three-card straight does not appear in the table, and thus is not worth holding unless it is suited (and hence a three-card straight flush; line 16) or consists of a jack, queen, and king (line 20). Similarly, the table never directs the player to hold a three-card flush unless it's part of a three-card royal or straight flush.

More significantly, the table never advises a player to hold a *kicker*: an unpaired high card held alongside a pair. If you have two 10s, a 5, a 2, and an ace of various suits, you should hold only the 10s and draw three cards, resisting the temptation to hold the ace as well, because holding onto it will not, in the long run, improve your position. This, of course, only applies to a game using Table 4.4 as its paytable. It is important when playing video poker to align your strategy with the rules and payoffs for the game you are playing. Nonetheless, holding a kicker is almost certainly going to decrease your long-term expectation, by as much as 4%, regardless of the exact game you're playing [35].

Example 4.2.31. What does the strategy chart say about the following hand?

$$5\clubsuit \ 8\heartsuit \ 4\diamondsuit \ 9\clubsuit \ 7\clubsuit$$

This hand contains the following:

- A three-card straight.

- A three-card inside straight flush.

- Two different four-card inside straights, although neither one contains any high cards.

Referring to Table 4.5, we see that line 16, "3-card straight flush," is the first one that matches our holdings, and so the correct course of action is to discard the $8\heartsuit$ and $4\diamondsuit$ and hope to complete the straight flush by drawing the $6\clubsuit$ and $8\clubsuit$. While the probability of pulling this off is only 1/1081, it' the option with the highest expected value.

It should be noted, of course, that there are multiple ways to win in this example; we might complete a flush or straight instead of a straight flush, or draw into two pairs or three of a kind.

The strategy table does not distinguish between inside and outside straight flushes because the payoff justifies the reach. If you hold three cards of a straight flush, the number of pairs of cards that will complete it ranges from one to three: $7\spadesuit 8\spadesuit 9\spadesuit$ can be filled out to a straight flush with either the 5 and 6, 5 and 10, or 10 and jack of spades, while $4\diamondsuit 6\diamondsuit 8\diamondsuit$ can only be completed to a straight flush with the 5 and 7 of diamonds. (A double inside straight flush such as $4\diamondsuit 6\diamondsuit 8\diamondsuit$ is sometimes called a *kangaroo straight.*) In any event, it's worth drawing two cards in search of the 50 for 1 payoff.

By making some small modifications to the paytable, we can arrive at Table 4.6, which is an 8/8 game with a payback percentage of 101.1% with perfect play [35, p. 125].

TABLE 4.6: Video poker paytable with a 101.1% return

Poker Hand	Payoff: 1 coin	Payoff: 5 coins
Royal flush	250	4700
Straight flush	200	1000
Four of a kind	40	200
Full house	8	40
Flush	8	40
Straight	8	40
Three of a kind	3	15
Two pair	1	5
Pair of jacks or higher	1	5

Of course, "perfect play" on this machine would be different from perfect play on the previous paytable. In particular, the sharply increased payoff for straight flushes means that perfect play calls for a gambler to pursue those more aggressively, even to the point where "Two-card royal flush" and "Two-card straight flush," which have no standing in Table 4.5, appear as entries in the corresponding strategy table [35].

Caribbean Stud Poker

Caribbean stud poker (CSP) is a table game in which players individually pit their five-card poker hands against the dealer's. While multiple players may bet against a single dealer hand, players are not in competition with one another. As an even-up game would hold no advantage for a casino offering this game, the CSP rules have been set to give the house an advantage.

Play begins with each player making an initial *ante* bet. Each player is dealt five cards from a single deck, and five cards are dealt to the dealer. No further cards will be dealt. One dealer card—the *upcard*—is exposed to the players. Based on the strength of their hands and the dealer's upcard, players may either fold, forfeiting their ante, or make an additional *call* bet that their hand will beat the dealer's. The call bet must be double the player's ante bet. This decision is the lone place where skill becomes important; unlike in standard poker, everything else in CSP is a matter of chance.

Example 4.2.32. You have made a $1 ante bet and are dealt the following CSP hand:

$$J\clubsuit \ 10\diamondsuit \ 8\spadesuit \ 8\heartsuit \ 2\diamondsuit.$$

The dealer's upcard is the $K\heartsuit$. The question you now face is whether or not a pair of 8s is likely to beat a qualifying dealer hand.

At this point, the game is essentially even. The house derives its advantage from how the hands are played out. The dealer's hand is said to *qualify* is least ace-king high. If the dealer fails to qualify, the players who have not folded are paid even money only on their ante bets, and the call bets push. If the dealer qualifies, the hands are compared, and if the player's hand beats the dealer's, the ante bet is paid at 1 to 1 and the call bet is paid in accordance with Table 4.7.

Example 4.2.33. Continuing with Example 4.2.32: If you make the call bet and the dealer fails to qualify, you win $1 on your ante bet, and the call bet is a push. If you make the call bet and the dealer qualifies, you win $2 on the call bet and $1 on the ante if his or her hand cannot beat your pair of 8s. If you make the call bet and the dealer beats your pair, you lose $3, and if you fold instead of calling, you lose $1.

The dealer's hand turns out to be

$$K\heartsuit \ K\diamondsuit \ Q\spadesuit \ Q\clubsuit \ 7\heartsuit,$$

TABLE 4.7: Caribbean stud poker payoffs

Player Hand	Payoff odds
Royal flush	100 to 1
Straight flush	50 to 1
Four of a kind	20 to 1
Full house	7 to 1
Flush	5 to 1
Straight	4 to 1
Three of a kind	3 to 1
Two pairs	2 to 1
One pair	1 to 1
AK high	1 to 1

so the dealer qualifies, and if you made the call bet, you were beaten by the dealer's two-pair hand.

There are two ways in which the casino gains an edge over the players:

- By requiring that the dealer have a qualifying hand, some potential high-paying player hands win only the ante bet if the dealer doesn't qualify. A player making a $10 ante bet and backing it up with a $20 call bet upon being dealt a flush, for example, wins only $10 instead of $ if the dealer's hand doesn't qualify. If you beat the low probability of .198% by receiving a dealt flush, you want to cash in for as much you can, but the requirement that the dealer qualify takes some of that opportunity away.

  This rule shows the true challenge facing a CSP player. You lose if the dealer's hand is good enough to beat yours, as in traditional stud poker, but you also lose, in the sense of winning less than you ought to, if the dealer has a bad hand.

- Players might fold winning hands based on the strength of the dealer's exposed card. For example, if the dealer shows an ace, a player might fold a low pair or ace-king hand that could beat a qualifying dealer hand with no pair or could win the ante bet if the dealer fails to qualify even with that ace. Part of the house advantage derives from the rule that players must make the call or fold decision before the dealer's hand is fully exposed.

Taken together, these factors mean that the house advantage in CSP starts at 5.22% with perfect play [25]. Imperfect play, of course, causes this advantage to rise.

Example 4.2.34. There are 1,135,260 possible dealer hands that do not qualify [27], so the probability of a dealer qualifying hand is

$$\frac{1,463,700}{2,598,960} \approx .5632.$$

Combining this with the known probabilities for poker hands leads to an interesting observation: even if you are dealt a rare high-ranking hand, you still have only a 56.32% chance of collecting on your call bet—and that assumes that you beat the dealer's qualifying hand.

With this probability in mind, we can develop an optimal player strategy, one that takes the form of "Fold all hands ranking lower than X and call with any hand that beats X." Clearly we want to call on any hand containing a pair or better, even allowing for the possibility that the dealer may beat our hand. This is because our losses are limited by the amount we've bet, while we may win more than we wager if our hand is good. Along the way, our strategy needs to consider the information provided by the dealer's upcard—if an ace is showing, it is somewhat more likely that the dealer will qualify than if the exposed card is a 4.

The question comes when we consider the Caribbean stud poker equivalent of bluffing—making the call bet on a weak hand in the hopes that the dealer won't qualify and we'll be paid on our ante bet. A very simple plan, one that barely counts as a "strategy," is simply to call on every hand, no matter what. This takes the element of skill completely out of the game and reduces CSP to a simple game of chance. Casinos love such a player.

We begin by considering the expectation of a $1 ante bet when we call despite holding a nonqualifying hand. We know that if we fold a nonqualifying hand, our expected return is –$1; it is to this value that we will compare our results. Just as the dealer has approximately a 44% chance of not qualifying, the probability that our hand is less than ace-king high is also .4368. Under these circumstances, we lose $3 every time we call and the dealer qualifies. If the dealer doesn't qualify, we win $1 each time regardless of the relative value of the two hands. It follows that the expected value of calling on a nonqualifying hand is

$$E = (-3) \cdot \left(\frac{1,463,700}{2,598,960}\right) + (1) \cdot \left(\frac{1,135,260}{2,598,960}\right) = -\frac{114}{91} \approx -\$1.25,$$

which is worse than simply folding and taking the $1 loss. However, if our hand contains an ace and a king but no pair, and thus has the potential to beat some qualifying dealer hands, then our expectation is higher than –$1.25 and is a function of the strength of the hand. The challenge comes in determining the cutoff point where $E > -1$. This calculation gets complicated because of the different payoffs for different winning player hands.

We begin by considering the best no-pair scenario, which eliminates the need to consider different payoffs. In [27], John Haigh carries out the calculations for calling on the highest possible hand without at least a pair: a no-flush

AKQJ9. On this hand, you will beat any qualifying dealer ace-king hand (we ignore the very small probability of a push, which depends on the exact cards you hold due to the need to consider flushes, and is less than 1/10,000). Since there are 167,820 five-card ace-king hands that are not straights or flushes [27, p. 326], it follows that your expectation is

$$E = \underbrace{(-3) \cdot \left(\frac{1,295,880}{2,598,960}\right)}_{\text{Dealer hand is a pair or better}} + \underbrace{(3) \cdot \left(\frac{167,820}{2,598,960}\right)}_{\text{Dealer hand is AK}} + \underbrace{(1) \cdot \left(\frac{1,135,260}{2,598,960}\right)}_{\text{Dealer fails to qualify}}$$

$$= -\frac{18,741}{21,658} \approx -\$.865.$$

Since the expected value of –$.865 is greater than –$1, it follows that if you hold this hand, a call bet is in order. The same holds true—as we guessed above—for any hand that beats AKQJ9.

At this point, our CSP strategy looks like this:

- **Call with any hand ranked AKQJ9 or higher.** This includes all hands that contain a pair or higher, and thus have the potential to beat a qualifying dealer hand.

- **Fold any hand not holding at least an ace and a king.** These are the hands that cannot beat a qualifying hand, and thus have an expectation of –$1 if folded vs. –$1.25 if called.

This covers over 93.4% of all possible dealt hands; the ace-king hands are all that remain. Once again referring to [27], we note that if we raise on AKQJ9, we gain about 13% relative to folding, whereas if we raise with the weakest ace-king hand, which would be AK432, we would lose about 25% versus folding. The conclusion is that the hand X mentioned above is about 2/3 of the way along the list of ace-king hands ranked from lowest to highest, which turns out to be AKJ82.

This analysis has not yet made use of the dealer's upcard, nor has it needed to. Haigh goes on to incorporate that information into the strategy, which concludes with the following rules determined with an eye on the possible number of dealer flushes:

- **Call on any hand that beats AKJ82, and fold any hand that doesn't.**

- **Call on AKJ82 if your hand includes all four suits, otherwise fold.** The four-suit requirement limits the number of possible dea flushes. [27, p. 215]

A more detailed analysis of CSP appears in [25], in which the simplest strategy of three presented simply calls for the player to call with any hand at AKJ83 (the so-called *beacon hand*) or better, and to fold otherwise—this is little different from the strategy above, developed three years later.

It must be noted here that knowledge of the cards held by the other players can be of great use to a CSP player. Since these cards cannot be among the dealer's hole cards, their presence in players' hands may provide important information about the dealer's chance of qualifying.

Example 4.2.35. Suppose that you are playing CSP with three other gamblers. You hold the nonqualifying hand $A\diamondsuit\ J\heartsuit\ 10\spadesuit\ 7\heartsuit\ 2\spadesuit$ and the dealer's upcard is the $Q\diamondsuit$. If you are informed that the other three queens are all distributed among your fellow gamblers' hands, you know that the dealer cannot hold a pair of queens, and thus the chance of him or her qualifying is diminished—it may then be worthwhile to call on your weak hand in the hopes of winning your ante bet.

For this reason, casinos are vigilant about not allowing players to communicate the contents of their hands to each other. At the Venetian and Palazzo casinos in Las Vegas, signs are posted at CSP tables stating that players revealing their hands to other players by showing the cards or talking about their holdings will have their hands ruled "dead," or out of play.

Let It Ride

Newer table games, including Caribbean stud poker, are often lumped together in a category called *carnival games*, which is a catch-all term describing games of more recent vintage, often possessing much higher house advantages than the three classic table games of roulette, craps, and blackjack. Some of these games appear in casinos for a brief time before disappearing due to lack of player interest; only a select few manage to overcome the obstacles facing game developers and find a lasting place on casino floors.

Another successful carnival game is Let It Ride (LIR), invented in 1993 by John Breeding. Breeding is the founder of Shuffle Master, a manufacturer of automatic card shuffling machines, and invented Let It Ride in the hopes of creating a new market for his company's product [3]. Let It Ride is a variation on poker, with the twist that players may choose to pull back some of their wagers if their hand looks unlikely to win.

The object of LIR is to make the best possible five-card poker hand from three personal cards and two community cards. As in CSP, winning hands are paid off at a fixed rate. Payoff tables vary among casinos; one common set of payoffs, and the one used at the Soaring Eagle Casino, may be found in Table 4.8.

The payoffs for LIR hands are equal to or better than those of CSP, except for the fact that LIR hands must contain at least a pair of 10s to win. Additionally, LIR does not contain the CSP requirement that the dealer have a qualifying hand—indeed, the dealer never receives a full five-card hand in Let It Ride.

Players begin a round of LIR by making three bets of equal amounts, in spaces labeled 1, 2, and $. The "$" bet may never be withdrawn, and remains

TABLE 4.8: Let It Ride payoffs [3]

Player hand	Payoff odds
Royal flush	1000 to 1
Straight flush	200 to 1
Four of a kind	50 to 1
Full house	11 to 1
Flush	8 to 1
Straight	5 to 1
Three of a kind	3 to 1
Two pairs	2 to 1
Pair of 10s or higher	1 to 1

in play until the hands are complete. Three cards are dealt face down to each player, and the dealer receives two cards dealt face down, which will serve as community cards for all players, as in Texas Hold'em. Upon examining their first three cards, the players may withdraw the "1" bet.

The dealer then exposes one of the community cards, and again, players may pull back one bet, this time the "2" bet. Finally, the second community card is revealed, and at this point, all remaining bets are resolved and paid off in accordance with the payoff table. If the player's hand does not contain at least a pair of 10s, then all bets left on the board lose and are collected.

Example 4.2.36. In a two-player game of Let It Ride, player 1 is dealt 9♠ 6♢ 5♡ and player 2 is dealt K♣ 8♣ 4♡. The first community card is the K♢, and the second is the J♡, giving player 1 a king-high hand and player 2 a pair of kings.

Once again, the calculated house advantage for this game is based on perfect player strategy, and any deviation from that strategy serves to increase the HA. The house edge, as in roulette, comes from paying players less than true odds on winning hands—far less, in some cases.

Example 4.2.37. The probability of drawing a straight flush in five cards is

$$\frac{36}{2,598,960} \approx \frac{1}{72,193}.$$

The casino pays off at 200 to 1 on this 72,192 to 1 shot, giving the house an immense edge.

How should a player make the all-important decisions whether or not to let bets 1 and 2 ride? There are a couple of obvious informal rules to follow: if you are dealt at least a pair of 10's, let them ride, but if your first four cards (including the first community card) are all 9s or lower, pull back bet #2. Indeed, pulling back bet #1 if you are dealt three nonsuited low cards is probably advisable.

But what about letting bets ride on three or four cards to a straight or flush? As a general rule, this is inadvisable, as the following strategy will show. Once again, computer simulation of many hands leads us in the best possible direction. By playing according to the following strategy, you can cut the house advantage to 3.5%, better than American roulette, although still not as good as blackjack or a pass line bet at craps [3].

- Pull back bet #1 unless you hold one of the following:

    - Any winning hand (a pair of 10s or higher).
    - Three cards to a royal flush.
    - Three cards to a straight flush with the lowest card at least a 3

    The restriction on this third option is to allow the straight to fill from either end—a hand such as $2\heartsuit\ 3\heartsuit\ 4\heartsuit$ can only be completed to a straight in two ways (ace-5 and 5-6), whereas $3\heartsuit\ 4\heartsuit\ 5\heartsuit$ admits three paths to a straight (ace-2, 2-6, and 6-7).

- After the first community card is exposed, pull back bet #2 unless you hold one of the following:

    - Any winning hand.
    - Four cards to a flush (including a royal flush).
    - Four cards to an open-ended straight. This is a straight that does not include an ace, and thus can be completed by two different ranks of card.
    - Four cards to an inside straight, provided that the lowest card is at least a 10—for example, any 10JKA. While the inside straigh is less likely to complete on the fifth card, the possibility of pairing one of your four cards to a winning pair makes up for the decrea probability of finishing the straight.

This strategy suggests rather clearly that the majority of starting hands do not merit letting the retractable bets ride. Low pairs do not justify a bet, nor do single high cards or three-card straights or flushes. Since you can never withdraw the $ bet, you are still in a position to win some money if the second community card, for example, completes a hand consisting of a pair of 7s and a 9 to two pairs or three of a kind.

Example 4.2.38. Suppose you are dealt $J\clubsuit\ Q\clubsuit\ 5\heartsuit$ playing one-on-one against the dealer. What is the probability that the two community cards will give you a winning hand?

There are two ways that you can win with this starting hand:

1. Pull additional jacks or queens in the community cards.

2. Find a pair of 5s, 10s, kings, or aces in the community cards.

In case 1, you need to pair either the jack or queen, and you have two cards to do so. Six cards will give you a winning combination. There are $_6C_1 \cdot {}_{44}C_1 + {}_6C_2 = 279$ ways for that to happen. In this total, the first term includes pairing the jack or the queen individually and includes the case of drawing a jack or queen with an additional 5, and the second covers the case where the community cards include two jacks, two queens, or one of each, which complete your hand to two pairs or three of a kind.

For case 2, there are $_4C_2 = 6$ ways to draw two of any rank, thus 24 ways for the community card to form a pair that gives you a winning hand: either three 5s or a pair of 10s, kings, or aces.

Putting everything together gives the following probability of winning:

$$\frac{279 + 24}{_{49}C_2} = \frac{303}{1176} \approx .2577,$$

which is not enough to justify risking two bets.

When playing any table game in a casino, it is important to be aware of the local rules, which are frequently available on the game's *rack card*: a card or brochure available in casinos that describes the rules of a game, including any local rules for player conduct. The Soaring Eagle's rack card for LIR includes the following rule:

> $25,000.00 Aggregate payout per round (This does not include BONUS bet)

Bonus bets will be examined in Chapter 5. How does this rule affect game play?

The casino has announced that on any given hand, a maximum of $25,000 will be paid out to all players. The corresponding player strategy is that you should limit your bets so that a big payoff will not be reduced by this limit. If you make three bets of $10 each and draw a royal flush (admittedly unlikely) with all three bets riding, then the casino rules state that you will only be paid $25,000, not the $30,000 called for by Table 4.8. Limiting your bets to a maximum of $8 each avoids this limit; a royal flush will then pay off $24,000, safely under the limit. As a bonus of sorts, you will lose money less rapidly by making smaller bets.

Three Card Poker

If five-card poker hands are too complicated, perhaps restricting your hand to three cards might be more to your liking. Apparently enough people feel this way to make *Three Card Poker* (3CP) another carnival game that's found a place in the game lineup at a number of casinos.

The title gives the key to the game: players are dealt three cards apiece, and match their cards up against the dealer's hand—once again, this is a game where players face off against the dealer instead of each other. 3CP is very much like Caribbean Stud Poker in that players make an Ante bet and then can choose, based on their cards, whether or not to fold or make a second "Play" bet. Also coming over from CSP is the notion of a qualifying hand for the dealer, which here is any hand holding at least a queen. One difference between CSP and 3CP is that none of the dealer's cards is exposed to help players make the decision whether or not to fold. If the dealer fails to qualify, Ante bets pay even money and Play bets push. If the dealer qualifies, winning player hands pay even money on both the Ante and Play bets; if the dealer's hand is higher, then both bets lose. Additionally, the Ante bet pays a bonus if the player's hand is a straight or higher, even if it is beaten by the dealer.

Example 4.2.39. What is the probability of the dealer qualifying?

There are $_{52}C_3 = 22,100$ three-card hands, and $_{40}C_3 = 9880$ of them contain no queens, kings, or aces. From this latter number, of course, we must remove any hand containing a pair or higher.

For example: Any card from 4 to J can be the highest card in a three-card straight flush; these hands run from 432 through J109. There are 8 ranks and 4 suits to consider, for a total of 32 such hands.

These "exceptional" hands are counted in Table 4.9.

TABLE 4.9: Three Card Poker: Qualifying hands with no AKQ

Hand	Count
Straight flush, J109 or lower	32
Three of a kind, jacks or lower	40
Flush, jack-high or lower	$4 \cdot {}_{10}C_3 - 32 = 448$
Straight, J109 or lower	$32 \cdot 4 \cdot 4 - 32 = 480$
Pair, jacks or lower	$10 \cdot {}_4C_2 \cdot 36 = 2160$

Adding up in this table gives 3160 possible qualifying hands with no card higher than a jack, and subtracting gives 6720 nonqualifying hands. The probability that the dealer qualifies is therefore

$$p = 1 - \frac{6720}{22,100} = \frac{15,380}{22,100} \approx .6959,$$

so the dealer qualifies about 70% of the time.

Armed with this information, we look next at the paytable for player hands that beat a qualifying dealer hand. The Ante Bonus paytable varies among casinos [73]; a common payoff structure is 1 to 1 on straights, 4 to 1 on three of a kinds, and 5 to 1 on straight flushes. Some casinos recognize a "Mini Royal Flush": a suited AKQ, and reward it with a 50 to 1 Ante Bonus payoff.

There is no paytable for the Play bet—all bets are paid at 1 to 1—b there is an optional Pair Plus bet that pays off player hands of at least a pair at odds. This bet must be made with the Ante bet before the cards are dealt, of course. Paytables can vary among casinos; one of the most common is Table 4.10. It should be noted that the Pair Plus bet is paid off even if the player's hand does not beat the dealer's or if the dealer fails to qualify.

TABLE 4.10: Three Card Poker payoffs: Pair Plus bet [73]

Hand	Payoff
Straight flush	40 to 1
Three of a kind	30 to 1
Straight	5 to 1
Flush	4 to 1
Pair	1 to 1

As we might expect, these payoffs are far below the probabilities of the various hands. In the case of a straight flush, there are 48 possible hands (any card except a deuce can be the high card in a three-card straight flush), and so $P(\text{Straight flush}) = 48/22,100 \approx 1/460.417$.

When considering CSP, we identified the lowest hand justifying a call bet, AKJ83. A similar analysis can be applied to 3CP; once again we seek the minimum hand for which the expectation is greater than the −$1 outcome achieved by folding and forfeiting the Ante bet. The "beacon hand" for 3CP turns out to be Q64: any player hand better than this should be backed with a Play bet and any lesser hand should be folded. Following this one piece of advice limits the casino's advantage to a mere 3.37%—actually quite reasonable for a table game. If instead you choose to "mimic the dealer" by calling on every hand holding at least a queen, that gives the casino a 3.45% edge—not much of a change. The nonstrategy of always calling, on the other hand, raises the HA to 7.65%: more than double that when using the beacon hand strategy above [73].

Pai Gow Poker

Pai Gow poker (PGP) is a variation on poker that uses seven cards per player and whose roots are found in *pai gow*, a domino-like game that is said to flourish in underground Chinese casinos. The best hand in pai gow is a 9, and so the name of the game is Cantonese for "make nine." In PGP, "make nine" is sometimes taken in the opposite way: to express a wish that the dealer will have the worst possible hand: 2345789 with no flush. An unusual feature of PGP is that players can serve as the banker and cover all other bets, including a bet against the casino's hand. All that is necessary is that the player be willing to bank and have enough money to pay off all other bets.

PGP uses a 53-card deck, including a single joker. This joker is not completely wild: it can only be used as an ace or to fill out a straight or flush; in this limited role, a wild joker is sometimes called a *bug*.

Example 4.2.40. With this restriction on the joker, a hand such as $Q\heartsuit$ $Q\diamondsuit$ Joker 5♠ 5♣ cannot be raised to a full house, since the bug cannot be used as a queen or 5 here. The joker is interpreted as an ace.

As with CSP, the game is played against a dealer hand. Players and the dealer are each dealt seven cards, which then must be "set," or sorted, into a five-card poker hand, the *back* hand, and a two-card hand, the *front* hand. It is a requirement of the game that the five-card hand be better than the two-card hand.

The back hand is ranked according to Table 2.4, with the following exceptions:

- Five of a kind is the highest hand, beating a royal flush.

- In most casinos, A2345 (called a *wheel*) is the highest straight flush and second highest straight, losing only to AKQJ10.

For the front hand, the highest hand is a pair of aces, followed by lesser pairs and then by high-card hands. Straights and flushes have no standing in the front hand.

Example 4.2.41. Suppose you are dealt the following PGP hand:

$$5♣\ 2\diamondsuit\ 7♣\ 9\heartsuit\ 2\heartsuit\ 8\heartsuit\ 6\heartsuit.$$

This hand can be split into a five-card straight and a pair of deuces, thus meeting the requirement that the back hand outrank the front hand.

Players make a single wager, which pays off at 19 to 20 (the casino charges a 5% commission on winning hands, as in baccarat) if both player hands beat the corresponding dealer hands. If one player hand beats the dealer and the other doesn't, the hand is a push. Ties on either hand, known as "copies," go to the bank, and hence the player loses. If a player is banking, then that player wins tie hands, which is some incentive for a player to bank. It is in these two ways that the casino derives its advantage, which comes to 1.57% from charging commissions and 1.27% from taking all ties, a total of 2.84% [28, p. 149–50].

The dealer has a fixed strategy, the *house way*, when setting his or her hands. This is similar to the rules guiding blackjack dealers: their actions are prescribed in advance and are independent of the players' hands. The house way is also the optimal strategy for players. While the particulars of the house way may vary slightly from casino to casino, the small differences have little effect on the overall house edge. One common version of the house way may be found in Table 4.11.

Inherent in the house way and the play of PGP hands is that making decisions about how to sort one's seven cards into front and back hands takes time, and so PGP proceeds somewhat more slowly than other table games such as blackjack and baccarat. While the edge is still with the casino, the

TABLE 4.11: Pai Gow Poker: House Way of setting hands [52]

Dealer's hand contains	Action
No pair	Place the highest card in the back hand and the next two highest cards in the front hand.
One Pair	Place the pair in the back hand and the next two highest cards in the front hand.
Two Pairs	Split the pairs if holding a pair of aces, or if holding a pair of face cards and 6s or higher. With two pairs of 6s or less, put the two pairs in the back hand. With any other two pairs, split them unless the hand contains an ace, then play the ace in the front hand.
Three Pairs	Always play the highest pair in the front hand.
Three of a Kind	Play the three of a kind in the back hand, except break up three aces.
Two Three of a Kinds	Play the lower three of a kind in the back hand and split the higher three of a kind.
5-card Straight	Keep the straight as the back hand.
6-card Straight	Use the highest card in the front hand.
5- or 6-card Straight with a Pair	Use the pair as the front hand.
Straight with 2 Pairs	Play according to the two-pair rules.
5-card Flush	Keep as the back hand.
6-card Flush	Use the highest card in the front hand.
5- or 6-card Flush with a Pair	Use the pair as the front hand.
Flush with 2 Pairs	Play according to the two-pair rules.
Straight & Flush, no Pair	Play the combination that results in the highest front hand.
Full House	Split the three of a kind into the back hand and the pair to the front hand. **Exception**: If you have a pair of 2s and an ace or king that can be played in the front hand, keep the full house together.
Four of a Kind	2–6: Always keep them together in the back hand. 7–10: Split unless a pair or ace and picture can be played in the front hand. J–K: Split unless also holding a pair of 10s or higher. Aces: Split unless a pair of 7s or higher can be played in the front hand.
Straight Flush	Keep as the back hand, with these exceptions: Split if holding two pair of 10s or higher, or a pair of aces and any other pair. If holding any other two pair with an ace, play two pairs as the front hand and the ace in the back hand. Play as a straight or flush instead of the straight flush if this puts a picture card or ace in the front hand.
Royal Flush	Keep as the back hand, with these exceptions: If also holding a pair, play the pair in the front hand. If also holding two pairs, play by the two-pair rules. Split the royal flush if an ace, king, or pair can be played in the front hand while retaining a straight or flush in the back hand.
Five Aces	Split unless a pair of kings can be played in the front hand.

slower game pace makes for fewer decisions per hour, and thus a PGP player will lose money more slowly than a gambler at quicker games.

Example 4.2.42. If the dealer holds the following hand:

$$5\clubsuit \ 6\heartsuit \ 5\diamondsuit \ 6\spadesuit \ 9\heartsuit \ Q\spadesuit \ J\clubsuit,$$

then the house way would keep the two pairs together since they're both 6s or lower, and then make the front hand as high as possible, so that the front hand would be $Q\spadesuit \ J\clubsuit$ and the back hand would be $5\clubsuit \ 5\diamondsuit \ 6\heartsuit \ 6\spadesuit \ 9\heartsuit$.

Example 4.2.43. If, in the hand in Example 4.2.42, the 5s were replaced by the corresponding aces, the house way would now specify that the aces be placed in the front hand and the 6s in the back hand, so we would have a front hand of $A\clubsuit \ A\diamondsuit$ and a back hand of $6\heartsuit \ 6\spadesuit \ Q\spadesuit \ J\clubsuit \ 9\heartsuit$.

Example 4.2.44. For the hand

$$J\heartsuit \ 9\diamondsuit \ 7\spadesuit \ K\heartsuit \ A\diamondsuit \ Q\diamondsuit \ 5\heartsuit,$$

the best that can be done is to make two high-card hands. Remembering that the back hand must outrank the front hand, the ace must go into the front hand. That having been done, it makes the most sense to put the king and queen into the back hand, making it as high as possible in the hopes of catching some low two-card player hands.

Example 4.2.45. What is the probability of being dealt a hand including the joker?

There are $_{53}C_7 = 154,143,080$ different seven-card hands that can be dealt. Of these, $_{52}C_6 = 20,358,520$ of them contain the joker—we simply count the number of ways to draw the remaining six cards from the 52 non-jokers. Accordingly, the probability of receiving the joker is

$$p = \frac{20,358,520}{154,143,080} = \frac{7}{53} \approx .1321,$$

or slightly less than once every seven hands.

Owing to the restrictions on how a wild joker may be used, there is only one five-of-a-kind hand in PGP: five aces. The probability of being dealt five aces is

$$p = \frac{_{48}C_2}{_{53}C_7} = \frac{1128}{154,143,080} \approx 7.318 \times 10^{-6},$$

where the numerator simply counts the number of ways to choose the other two cards from the 48 deuces through kings. Oddly, it's extremely unlikely that a skilled player who is dealt five aces will put them in the five-card hand. The house way stipulates that five aces should be split unless the other two cards are a pair of kings. The reason for this is easy to understand: a pair of aces is the best possible front hand, and moreover cannot be beaten or tied by the dealer in this case, since the player holds all of the other aces. Holding all of the aces also means that the dealer cannot beat a KK front hand. Three aces is a pretty strong back hand, and thus very likely to win without additional value.

Example 4.2.46. A hand that is available in PGP but not in any five-card poker game is two three-of-a-kinds. What is the probability of being dealt this hand?

We must split this into two cases: hands with and without the bug being used as a third ace. If you do not hold three aces, then the number of hands with two three-of-a-kinds is

$$N_1 = {}_{12}C_2 \cdot ({}_4C_3)^2 \cdot {}_{45}C_1 = 47,520.$$

In this expression, the first factor is the number of ways to pick two ranks that are not aces, the second counts the ways to select the cards, and the third accounts for the seventh, nonmatching card. The joker may be drawn as card

#7, since with two sets of three matching nonaces, it will not be used to fill out a straight or flush, and so will be counted as an ace.

If you hold three aces, then the number of hands is

$$N_2 = {}_{12}C_1 \cdot {}_4C_3 \cdot {}_5C_3 \cdot {}_{44}C_1 = 21,120.$$

Here, the first term selects the rank of the other three-of-a-kind, and the second term counts the number of ways to choose those cards. The factor is the number of ways to choose three aces from a set including the bug as a potential fifth ace, and the last term fills in the odd card, which cannot be a card already chosen or any of the three remaining cards—possibly including the joker—that would elevate this hand to "four of a kind plus three of a kind."

The probability of two three-of-a-kinds is therefore

$$p = \frac{N_1 + N_2}{{}_{53}C_7} = \frac{68,640}{154,143,080} \approx 4.453 \times 10^{-4}.$$

PGP allows a player to bank the game and play against the other players and the dealer, an opportunity which is customarily offered to each player in turn. It is reasonable to ask why this is allowed. At first glance, it would appear that a player banking the game would reap the advantage of the house edge working in his or her favor—but certainly no casino would allow that.

The advantages of banking are clear: the HA works in your favor rather than against you, the fact that ties go to the bank has the potential of reducing your losses, and you can win money from every other player. Of course, the big negative is that you might lose to many other players if you are dealt bad cards or if your skill at setting hands is inferior to theirs.

Sports Betting

Example 4.2.47. In Example 3.2.9, we looked at the daily parlay cards offered by Station Casinos. If we assume that your chance of picking the winner of each game you bet on is $\frac{1}{2}$, what is the expected value of a parlay card bet?

The expected value (EV) depends on how many games you select. In general, the expected value of a \$1 bet on an n-team parlay card is

$$E = (x-1) \cdot \left(\frac{1}{2}\right)^n + (-1) \cdot \left(1 - \left(\frac{1}{2}\right)^n\right) = x \cdot \left(\frac{1}{2}\right)^n - 1,$$

where x is the payoff, stated as "x for 1." Remember that when payoff odds are stated in this fashion, the payoff includes the amount of the original bet. For the payoff options offered on the parlay card, Table 4.12 shows the expected values and corresponding house advantages.

TABLE 4.12: Parlay card payoffs and expectations: $p = .50$

n	Payoff	Expected value	HA
3	6 for 1	−$.25	25%
4	12 for 1	−$.25	25%
5	23 for 1	−$.281	28.1%
6	45 for 1	−$.297	29.7%
7	80 for 1	−$.375	37.5%
8	160 for 1	−$.375	37.5%
9	320 for 1	−$.375	37.5%
10	800 for 1	−$.219	21.9%

Some things are immediately apparent from this table:

- The house advantage on every bet offered is *huge*—on a par with keno and among the highest of all the bets available in a casino.

- The parlay with the lowest HA—picking ten games correctly—is a the one with the lowest probability of success, $\dfrac{1}{1024}$.

- The HA for picking seven, eight, and nine games correctly is the same. This is because the payoff has doubled while the probability of winning has decreased by $\frac{1}{2}$, and these factors cancel out.

- If the pattern for seven, eight, and nine games had been continued through to ten games, the payoff would have been 640 for 1. Since the 800 for 1 payoff is printed on the parlay card in larger and bolder type than the others, we may conclude that the casino wishes to attract action on this betting option by offering a higher payoff, and that, given the low probability of winning this bet, they do so at little risk.

But what if you know something about sports and can pick winners at a better-than-50% rate? If you can do even as well as 53% on every game, you can cut the HA to a far more reasonable level—and even turn the game to your advantage for $n = 9$ or 10. The EV for a $1 bet then increases to

$$E = x \cdot (.53)^n - 1.$$

These values, for the various values of n, are collected in Table 4.13. A negative HA in this table represents a player advantage.

Here, then, is one case where skill plays a part in gambling and where, under the right circumstances, the game can be turned in the gambler's favor.

Turning now to betting on single games: If a casino sports book says that the Miami Dolphins are a $3\frac{1}{2}$-point favorite over the Jacksonville Jaguars, bets on Miami only pay off if the Dolphins win by at least 4 points; bets on

TABLE 4.13: Parlay card payoffs and expectations: $p = .53$

n	Payoff	Expected value	HA
3	6 for 1	−$.107	10.7%
4	12 for 1	−$.053	5.3%
5	23 for 1	−$.0382	3.82%
6	45 for 1	−$.0026	.26%
7	80 for 1	−$.0602	6.02%
8	160 for 1	−$.0038	.38%
9	320 for 1	$.0559	**−5.59%**
10	800 for 1	$.3991	**−39.91%**

Jacksonville win if the Jaguars win the game or lose by no more than 3. In effect, a team tabbed as an x-point underdog starts the game—for betting purposes—with x points. Half-point increments are often seen and eliminate the possibility of a wager ending in a tie.

In setting the point spreads for football games, casinos do so in order to encourage approximately equal action on both teams. The casino makes its money in the way that bets are accepted: gamblers wishing to bet on a game must bet $11 to win $10. This extra $1 is called the *vigorish*, or "vig" for short. The sports book collects $11 from every player and pays out $21 (the original $11 bet plus $10 in winnings) to approximately half of them, leaving the extra $1 vig from the losing bettors as the casino's profit.

Example 4.2.48. Assuming that the line has been properly set, players are essentially betting $11 on a coin toss hoping to win $10, and so the expectation is

$$E = (10) \cdot \left(\frac{1}{2}\right) + (-11) \cdot \left(\frac{1}{2}\right) = -\$\frac{1}{2},$$

which results in a house edge of $.5/11 \approx 4.545\%$, no matter which team the bettor chooses.

Of course, point spreads are sometimes known to shift in the days preceding a game, due to changing circumstances such as player injuries or a casino's desire to keep its books balanced with approximately equal action on each team. In our example, if too many people are betting on Miami giving 3 points, the casino may change the line to "Miami $-4\frac{1}{2}$," meaning that the Dolphins are now a $4\frac{1}{2}$-point favorite, with the intent of encouraging people to bet on Jacksonville and balance the amounts being wagered on each team.

This can allow an observant and quick-acting gambler the chance to make separate bets on both teams in a game and hope for a "straddle," where both bets win. For example, if someone has already placed a bet on Jacksonville getting $3\frac{1}{2}$ points above, and the line moves to make Miami only a $2\frac{1}{2}$-point favorite, making a second bet on Miami on these terms means that if the Dolphins win by exactly 3 points, both of these bets pay off.

The risk here is that in the event of any other outcome to the game, one bet wins and the other loses, for a net player loss of the $1 vig, so this may not be advisable if the line has only moved by one point.

An alternate approach to moving a point spread is offering a *money line* which requires that bettors wishing to wager on the favorite risk more money than if they were betting on the underdog. In the Miami-Jacksonville game above, rather than switching Miami from a $3\frac{1}{2}$- to a $4\frac{1}{2}$-point favorite, the bookmaker may reset the line to something like "−120/+110," which means that a gambler wishing to bet on the favorite (Miami) must risk $120 to win $100, still against a $3\frac{1}{2}$-point line, while someone betting $100 on the underdog (Jacksonville) would win $110 for a $100 bet. In this manner, moving the line substitutes for changing the point spread as a means to encourage approximately equal action on each team, and the casino avoids the prospect of being caught in a straddle, because the point spread never changes.

Example 4.2.49. In early 2013, some Las Vegas sports books were offering the following money line on the coin toss at Super Bowl XLVII:

$$\text{Heads: } -102$$
$$\text{Tails: } -102$$

Translated into betting activity, this meant that, in making this bet, you would be risking $102 on the outcome of the coin toss, hoping to win $100 (or some other allowable wager at the same odds: $51 hoping to win $50 or $ to win $200, for example). Since a coin toss is the ultimate 50/50 proposition, the expected value of this bet is

$$E = (100) \cdot \left(\frac{1}{2}\right) + (-102) \cdot \left(\frac{1}{2}\right) = -\$1,$$

and the corresponding house advantage is $\frac{1}{102} \approx .98\%$.

While this is a very low HA, it cannot escape notice that making this bet calls for risking more money than you stand to win—on the toss of a coin. (The coin came up heads.)

Punchboards

In considering the Ninety Percenter punchboard from Example 3.2.19, one is drawn to the fact that, due to the $5 prize for punching out the last spot, a gambler is guaranteed a profit if there are 49 or fewer unpunched spots: by purchasing and punching all of them. This profit is at least 10¢ and may be more if there are additional winning numbers still on the board. At the same time, purchasing all 800 spots on a fresh punchboard is a recipe for a guaranteed loss of $8 if the board is accurately named.

Example 4.2.50. Given this progression from negative to positive expectation, it's reasonable to ask the following question: At which number of unpunched spots is the expected return from purchasing them all a maximum? To assess this question, we need to develop an expression for the expected number of $1-winning spots left on the board after a certain number of punches have been sold. Let the number of previous punches be n, where $0 \leq n \leq 799$, and denote by N the expected number of winners drawn; we then have

$$N = \sum_{k=0}^{n} k \cdot P(k \text{ winners in } n \text{ punches}) = \sum_{k=0}^{n} k \cdot \left(\frac{{}_{67}C_k \cdot {}_{733}C_{n-k}}{{}_{800}C_n} \right).$$

N is a function only of n and may rightly be denoted $N(n)$ to show this functional relationship. Since the expression ${}_nC_r$ is 0 if $r > n$, this sum can be truncated with an upper limit of 67 if $n > 67$.

It follows that the expected number of $1 winners remaining after punches is $67 - N(n)$. If you go on to purchase the last $800 - n$ spots, your expected return as a function of n is

$$E(n) = \$\left(67 - N(n) + 5 - \frac{800 - n}{10} \right) = \$\left(\frac{n}{10} - N(n) - 8 \right).$$

This can be simplified considerably:

$$E(n) = \frac{13}{800}n - 8,$$

from which we can see that the expectation is linear and increasing, and first exceeds 0 when $n = 493$. It follows that a Ninety Percenter punchboard with 307 or fewer spots remaining has a positive expectation.

The *maximum* expectation is achieved with one spot left to be punched, although it is difficult to imagine a situation where a gambler would have encountered this punchboard with only one spot remaining. Since the return on the last spot is guaranteed to be at least $5, it is unlikely that another gambler would have walked away when $5 (or possibly $6, with probability 67/800) could be had for a dime.

4.3 BINOMIAL DISTRIBUTION

We begin by considering an example:

> *Suppose you make 100 $1 bets on red at an American roulette wheel. What is the probability that you will win on 60 of the spins?*

We might consider this problem by looking at your total winnings, should this occur—that would be $20. Trying to calculate this probability directly quickly leads us into an algebraic morass from which it is difficult to emerge with an answer that we can trust.

Solving this problem is facilitated by introducing the concept of a *binomial experiment*.

Definition 4.3.1. A *binomial* experiment has the following four characteristics:

1. The experiment consists of a fixed number of successive trials, denoted by n.

2. Each trial has exactly two outcomes, denoted *success* and *failure*.

   In practice, it is often possible to amalgamate multiple outcomes into a single category to get down to two. For example, in the roulette problem posed above, we can denote the 18 red numbers as "success" and the 20 nonred numbers as "failure"—if we lose our bet on red, it matters little whether the number that was actually spun was black or green, or what the number was.

3. The probabilities of success and failure are constant from trial to trial. We denote the probability of success by p and the probability of failure by q, where $q = 1 - p$.

4. The trials are independent.

Definition 4.3.2. A random variable X that counts the number of successes of a binomial experiment is called a *binomial* random variable. The values and p are called the *parameters* of X.

The experiment described in the example above meets the four listed criteria and is therefore a binomial experiment. If we change the experiment to "Begin betting on red on successive spins of a roulette wheel, and let the random variable X be the number of spins required to win exactly ten times," then the new experiment is not binomial. Since the number of trials is no longer fixed at the outset, criterion 1 is no longer true.

If X is a binomial random variable with parameters n and p, the formula for $P(X = r)$ can be derived through the following three-step process:

1. Select which r of the n trials are to be successes. This can be done in $_nC_r$ ways, as the order in which we select the successes does not matter.

   If we think of the trials as a row of n boxes, each to be designated "success" or "failure," what we're doing here is determining which r the n boxes are successes.

2. Compute the probability of these r trials resulting in successes. Since the trials are independent, this probability is p^r.

3. We must now ensure that there are *only* p successes. This is done by assigning the outcome "failure" to the remaining $n - r$ trials. The probability of this many failures is $(1 - p)^{n-r} = q^{n-r}$.

Multiplying these three factors together gives the following result, called the *binomial formula*:

Theorem 4.3.1. *If X is a binomial random variable with parameters n
p, then*

$$P(X = r) = {}_nC_r \cdot p^r \cdot q^{n-r} = {}_nC_r \cdot p^r \cdot (1-p)^{n-r}.$$

We can now revisit the example that started the section with this new
insight. The experiment described meets the four criteria of a binomial experiment, where the amalgamation described in the definition is used to get
down to two outcomes. The parameters of the experiment are $n = 100$ and
$p = \frac{18}{38}$, and so with $r = 60$, we find that

$$P(X = 60) = {}_{100}C_{60} \cdot \left(\frac{18}{38}\right)^{60} \cdot \left(\frac{20}{38}\right)^{40} \approx .0033.$$

Binomial probabilities can be computed directly on a TI-84+ calculator
by using the **binompdf** routine, which is found by pressing $\boxed{\text{2nd}}$ $\boxed{\text{DISTR}}$
and choosing option **A: binompdf**. The calculator will prompt you for the
values of n, p, and r. "pdf" is an abbreviation for "probability distribution
function."

Another calculator routine, **binomcdf**, computes binomial probabilities
involving inequalities. "cdf" stands for "cumulative distribution function" and
finds probabilities of the form $P(X \le n)$.

Example 4.3.1. In the roulette betting scenario described above, what is th
probability that you will be ahead after 100 bets?

Being ahead after 100 even-money bets corresponds to winning at least
51 bets, which may be interpreted probabilistically as $P(X \ge 51)$. Calculating this probability using the binomial formula would require 50 separate
calculations and one large sum, or some advanced estimation techniques. The
binomcdf routine, accessible as option **B: binomcdf** on the distribution
menu, computes $P(0 \le X \le r)$ when provided with values for n, p, and r.

For this problem, we need $P(X \ge 51)$, which is $1 - P(0 \le X \le$
by the Complement Rule. Using the TI-84+ shows that this probability is
approximately .2650.

Example 4.3.2. Suppose instead that you make 100 consecutive roulette
bets on the number 13. What is the probability that you will be ahead after
100 spins?

A single-number roulette bet pays off at 35 to 1, so if you win three times,
the $105 you win will offset your $97 in losses and leave you with $8 in profit.
To show a profit, then, you must win at least three times, so we need to find
$P(X \ge 3)$, or $1 - P(X \le 2)$. With $n = 100$ and $p = \frac{1}{38}$, we compute this on
the TI-84+ by finding $1 - P(X \le 2)$, which is .1649.

If a random variable is binomial, computing its expected value is simple.

Theorem 4.3.2. *If X is a binomial random variable with parameters n
p, then $E(X) = np$.*

Put simply, the average number of successes is the number of trials multiplied by the probability of success on a single trial.

Proof.

$$E(X) = \sum_{x=0}^{n} x \cdot P(X = x)$$

$$= \sum_{x=0}^{n} x \cdot {_nC_x} \cdot p^x \cdot q^{n-x}$$

$$= \sum_{x=0}^{n} x \cdot \frac{n!}{(n-x)! \cdot x!} \cdot p^x \cdot q^{n-x}.$$

Since the $x = 0$ term is equal to zero, we can drop that term from the sum and renumber starting at 1:

$$E(X) = \sum_{x=1}^{n} x \cdot \frac{n!}{(n-x)! \cdot x!} \cdot p^x \cdot q^{n-x}$$

$$= \sum_{x=1}^{n} \frac{n!}{(n-x)! \cdot (x-1)!} \cdot p^x \cdot q^{n-x}$$

$$= np \cdot \sum_{x=1}^{n} \frac{(n-1)!}{(n-x)! \cdot (x-1)!} \cdot p^{x-1} \cdot q^{n-x}$$

$$= np \cdot \sum_{x=1}^{n} \frac{(n-1)!}{[(n-1) - (x-1)]! \cdot (x-1)!} \cdot p^{x-1} \cdot q^{n-1-x+1}.$$

If we substitute $y = x - 1$ in this last sum, we have

$$E(X) = np \cdot \sum_{y=0}^{n-1} \frac{(n-1)!}{[(n-1) - y]! \cdot y!} \cdot p^y \cdot q^{n-1-y}$$

$$= np \cdot \sum_{y=0}^{n-1} P(Y = y).$$

This sum is the sum of all of the probabilities of a binomial random variable Y with parameters $n - 1$ and p, and so adds up to 1 (Axiom 2), completing the proof.

Example 4.3.3. In Example 4.3.1, the expected number of wins in 100 spins of the wheel is $E = np = 100 \cdot \frac{18}{38} \approx 47.37$.

It follows from this result that your average winnings after making 100 of these bets would be

$$(1) \cdot 47.37 + (-1) \cdot 52.63 = -\$5.26,$$

or 5.26% of the amount wagered—which is, of course, the long-term HA on a single bet.

Example 4.3.4. If we consider a bet on a single number rather than on red in American roulette (as in Example 4.3.1), the expected number of wins in 100 spins is $E = np = 100 \cdot \frac{1}{38} \approx 2.63$. Once again, your average winnings would be $-\$5.26$ after 100 \$1 bets.

Example 4.3.5. *Twenty-Six* was a popular game in the Midwest (in underground gambling operations, of course) in the 1950s [59]. To play Twenty-Six, you select a number from 1 to 6–called your *point*–and roll 130 six-sided dice. In practice, this was done by having the player roll ten dice 13 times. If your point comes up 26 times or more, you are paid off at 3 to 1 odds. If you roll your point 33 or more times, you are paid 7 to 1. At the other end of the scale, if you roll your point 11 or fewer times, that pays 3 to 1 also, and if you roll your point exactly 13 times, the payoff is 2 to 1.

Suppose that you choose the point 5. Before breaking this game down, we note that the mean number of 5s you can expect to roll in 130 tosses of a fair die is $130 \cdot \frac{1}{6} = 21\frac{2}{3}$, so the number of 5s that triggers a payoff is pretty far removed—in both directions—from the expected value. With that in mind, we note that there are four possible numerical outcomes for a \$1 bet:

- Win \$1 with 13 5s.

- Win \$3, either by rolling fewer than 11 5s or between 26 and 32 of the

- Win \$7, by rolling 33 or more 5s.

- Lose \$1, in every other outcome.

We shall treat these four cases separately. If we let X denote the number of 5s rolled in 130 trials, we have a binomial random variable with parameters $n = 130$ and $p = \frac{1}{6}$.

Winning \$1 is the easiest case to compute, since this involves only a single outcome.

$$P(X = 13) =_{130} C_{13} \cdot \left(\frac{1}{6}\right)^{13} \cdot \left(\frac{5}{6}\right)^{117} = .0109.$$

To win \$3, you must roll 11 or fewer 5s, or between 26 and 32 5s inclusiv Mathematically, we seek $P(X \leq 11) + P(26 \leq X \leq 32)$. On the TI-84+, this is easily computed with three applications of the **binomcdf** routine. $P(X \leq$ is available with a direct application of **binomcdf** and is .0053.

To compute the second term in the sum above, we need to find $P(26 \leq X \leq 32) = P(X \leq 32) - P(X \leq 25)$, which is done with **binomcdf**(130, 1/6, 32) - **binomcdf**(130, 1/6, 25). This probability is found to be .1746. Adding this to $P(X \leq 11)$ gives $P(\text{Win } \$3) = .1799$.

Winning $7 requires rolling 33 or more 5s. $P(X \geq 33) = 1 - P(X \leq 32) = 1 - .9925 = .0075$.

Losing $1 is the outcome in all other cases. By the Complement Rule, $P(\text{Lose } \$1) = 1 - .0109 - .1799 - .0075 = .8017$.

We then have the following probability distribution for Y, the number of dollars won:

y	−1	1	3	7
$P(Y = y)$	.8017	.0109	.1799	.0075

and the expected value of a $1 bet is then found to be -$.1986—a 19.86% advantage for the house.

Aside from the general rule about "gambling games always favor the house," it might be reasonable to ask if there is any way to identify particularly bad bets such as this one before risking money. One tool that may be useful is the *standard deviation* of a random variable, which is denoted by the Greek letter sigma: σ.

Definition 4.3.3. The *standard deviation* σ of a random variable X with mean μ is

$$\sigma = \sqrt{\sum [x^2 \cdot P(X = x)] - \mu^2}$$

where the sum is taken over all possible values of X.

Informally, σ is a measure of how far a typical value of X lies from the mean. Computing σ using the formula above is an arithmetically intense process:

- Compute the mean of the random variable.

- For each value of x that X can attain, multiply x^2 by $P(X = x)$ and add up the products.

- Subtract the square of the mean from this sum.

- Take the square root of the difference. Notice that the use of the square root guarantees that $\sigma \geq 0$.

Fortunately, it is seldom necessary to perform these calculations by hand, as calculators and computer software will readily compute σ. In the special case where the random variable is binomial, we have the following simple result:

Theorem 4.3.3. *If X is a binomial random variable with parameters n p, then the standard deviation of X is given by*

$$\sigma = \sqrt{np(1 - p)} = \sqrt{npq}.$$

While the expected value of a random variable tells us about where a "typical" value of X lies, the standard deviation gives us information about the "spread" of the values of X. The nature of the random variable X allows us to use the mean and standard deviation to derive useful information about the distribution of the data set. For example, in considering a suitably large number of tosses of two fair dice, the distribution of sums is approximately bell-shaped (many values near the mean, and fewer values as we move away from the mean of 7 in either direction) and symmetrically distributed about the mean. See Figure 4.2, which illustrates the results of 1000 simulated rolls of 2d6, for an example.

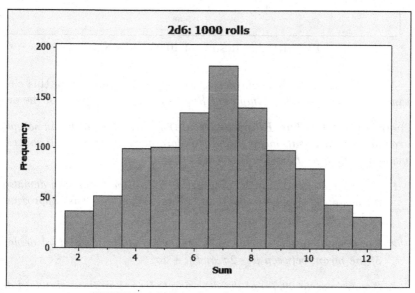

FIGURE 4.2: Results of 1000 rolls of 2d6

The symmetry of the graph becomes more pronounced as the number of rolls increases; Figure 4.3 shows a histogram of 10,000 rolls of 2d6, which is more symmetric than Figure 4.2.

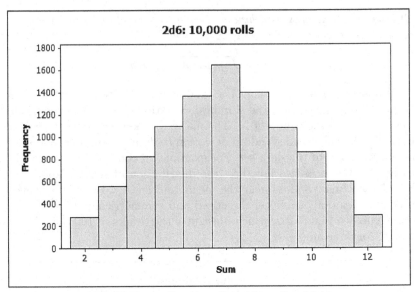

FIGURE 4.3: Results of 10,000 rolls of 2d6

Data sets that are distributed this way are called *normal*. In this circumstance, a result called the *Empirical Rule* is a good description of the data.

Theorem 4.3.4. *(**The Empirical Rule**) If the values of many samples of a random variable with mean μ and standard deviation σ are bell-shaped and symmetric, then we have the following result:*

1. *Approximately 68% of the data points lie within 1 standard deviation of the mean: between $\mu - \sigma$ and $\mu + \sigma$. (This is often stated as "approximately 2/3.")*

2. *Approximately 95% of the data points lie within 2 standard deviations of the mean: between $\mu - 2\sigma$ and $\mu + 2\sigma$.*

3. *Approximately 99.7% of the data points lie within three standard deviations of the mean: between $\mu - 3\sigma$ and $\mu + 3\sigma$. (This is often stated as "almost all.")*

For many practical purposes, an experimental result is deemed to be statistically significant if it is at least 2 standard deviations away from the mean, which means that its probability—under the assumption that there is no unusual effect and all of the deviation from the mean is due to random chance—is less than 5%. Since the random variable is symmetric, this 5% is evenly distributed between results greater than 2 SDs above μ and results less than 2 SDs below μ. This 2 SD standard is a convention agreed upon by the statistics community; it does not fall out of any equation as a rigorously derived standard.

By using 95% as a minimum, what we are saying is that 19 out of 20 times that we identify a result as due to something other than random variation, we will be correct, and this level of confidence is acceptable in many lines of inquiry. Some fields may have more exacting standards: in experimental particle physics, for example, the standard for confirming a discovery is "5 sigma," or at least 5 standard deviations away from the expected value, corresponding to $P(\text{Chance event}) < \dfrac{1}{3.5 \times 10^6}$.

Example 4.3.6. Reconsidering Twenty-Six in light of this new information, we can calculate the SD of the binomial random variable X:

$$\sigma = \sqrt{npq} = \sqrt{130 \cdot \frac{1}{6} \cdot \frac{5}{6}} \approx 4.25.$$

Figure 4.4 indicates that the number of 3s (for example; any number from 1 to 6 could be chosen) has a normal distribution, and so the Empirical Rule may be used to analyze this game.

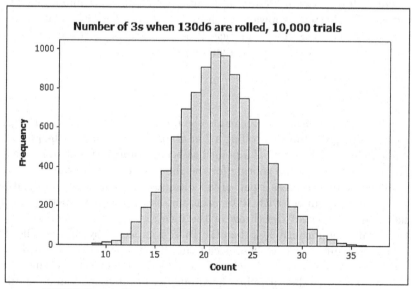

FIGURE 4.4: Number of 3s rolled in 10,000 rolls of 130d6

The Empirical Rule then tells us the following:

$$P(\mu - \sigma < X < \mu + \sigma) = P\left(16\frac{11}{12} < X < 25\frac{5}{12}\right) \approx .68$$

$$P(\mu - 2\sigma < X < \mu + 2\sigma) = P\left(13\frac{1}{6} < X < 30\frac{1}{6}\right) \approx .95$$

$$P(\mu - 3\sigma < X < \mu + 3\sigma) = P\left(8\frac{11}{12} < X < 34\frac{5}{12}\right) \approx .997.$$

To win money, we must either roll 26 or more of our point, 11 or fewer, or exactly 13. Since 13 is an exact value and not an interval, the Empirical Rule is not necessary to assess the likelihood of that result. With $\mu = 21\frac{2}{3}$ $\sigma = 4.25$, we can see that 11 is about 2.5 SDs below the mean, and so is outside the interval from 14 to 30 within 2 SDs of μ. The symmetry of the distribution tells us that $P(X \leq 13) \approx .025$, and this includes nonpaying values of 12 and 13 as well all money-winning values of 11 or less. The situation for 33 or more is the same; 33 lies about 2.67 standard deviations from the mean.

For rolling between 26 and 32 points, our chance of winning is certainly better—but "better" is always a relative term, and here it's relative to "less than a 2.5% chance." The Empirical Rule, by itself, suggests that this probability is less than 16%—hardly a winning proposition.

Craps

The binomial distribution can also be useful in analyzing certain craps situations. The probability of rolling a 7 at craps is $p = \frac{1}{6}$, and so the probability of not rolling a 7 is $q = \frac{5}{6}$. If a point has been established and you have a bet on the pass line, you are rooting for the shooter not to throw a 7.

While no one can toss a pair of fair dice under casino conditions and bring up the desired number with 100% accuracy, there are dice experts who claim the ability to control the throw of the dice so that the proportion of 7s is statistically significantly less than the 16.67% expected by chance. The stakes are not insignificant: If a shooter can cut the proportion of 7s rolled from 1 in 6 to 1 in 7, he or she can achieve an edge of 4.4% over the casino [102, p. 279]. Since the number of 7s in a fixed number of tosses is a binomial random variable, we have the mathematical model to assess these shooters' skill level.

A controlled shooter's regimen has two components: *setting* the dice and *tossing* them. Setting the dice refers to how the dice are aligned in the shooter's hand, relative both to each other and to his or her fingers. In a common set intended to throw hardway combinations, the dice are held together as a rectangular solid, arranged with hardway pairs from 2-2 through 5-5 together on the outside and the 1 and 6 faces on the vertical sides. The dice, when tossed, rotate about an axis through the 1 and 6 faces, and if thrown properly, turn up more hardways and fewer 7s than expected by random chance, since the 1-6 combination is less likely [101].

A controlled toss is intended to minimize, if not totally eliminate, random bouncing of the dice. At the craps table, it is expected that the dice will rebound off the far wall of the table, which is covered with pyramid-textured rubber to induce random rolling, before coming to rest. In an ideal controlled toss, the dice leave the shooter's hand together, fly through the air touching or nearly touching with some backspin, bounce off the table and hit the wall nearly together, and finally drop back to the table as gently as possible, without much random effect imparted from the wall.

Suppose that a controlled shooter claims the ability to throw fewer 7s than the 1 in 6 that would be expected from throwing the dice at random. How would we assess their performance, ideally before risking a lot of money based on their claim? How many fewer 7s than expected would be enough to convince us that there's a real effect here, and not just random fluctuation?

Example 4.3.7. In [101], Stanford Wong describes an experiment testing his and another shooter's ability to toss fewer 7s than expected by chance. In 500 rolls, the two men tossed only 74 7s. Does this constitute a statistically significant result?

The proportion of 7s in this experiment was $74/500 = .148$, certainly less than $1/6$. Since the random variable is binomial, we can use Theorems 4.3.2 and 4.3.3 to compute the mean and standard deviation.

$$\mu = np = 500 \cdot \frac{1}{6} = 83.333.$$

$$\sigma = \sqrt{npq} = \sqrt{500 \cdot \frac{1}{6} \cdot \frac{5}{6}} = 8.333.$$

It follows that the probability of tossing fewer than

$$\mu - 2\sigma = 83.333 - 2 \cdot 8.333 = 66.667$$

7's is .025, so a result of 74, while considerably less than 83.333, does not quite rise to the level of statistical significance.

In devising this challenge, Wong was aiming at a slightly lower level of significance, and set his target number at 79, roughly half a standard deviation below the expected number. The probability of tossing 79 or fewer 7s in 500 tosses is about .37—not statistically significant, but perhaps meaningful enough to give an advantage to gamblers betting on the pass line with him throwing the dice [101]. Recall that the HA for a bet on the pass line is only 1.41%, so it wouldn't take a large deviation from random chance to tip the edge over to the players.

Roulette

Controlled dice shooting is not illegal. Neither is *wheel clocking*, or repeatedly observing and recording the successive spins of a roulette wheel in an effort to detect which numbers, if any, the wheel favors through simple mechanical slackness. While professional-grade roulette wheels are very carefully machined to ensure totally random results, they are nonetheless physical objects and so subject to mechanical wear and tear that could potentially lead to some numbers appearing more often than random chance would suggest. Casino-quality roulette wheels cost thousands of dollars, and so are not replaced as often as cards and dice are.

How often that "more often" is can be determined through the use of the binomial formula together with the Empirical Rule. To do this effectively, it is necessary to consider a fairly large number of spins, since the probability of spinning a given single number is only $p = \frac{1}{38}$.

Example 4.3.8. If you clock a wheel through 200 spins, the mean and standard deviation of the number of times any one number appears are

$$\mu = np = 200 \cdot \frac{1}{38} = \frac{200}{38} \approx 5.263$$

and

$$\sigma = \sqrt{npq} = \sqrt{200 \cdot \frac{1}{38} \cdot \frac{37}{38}} \approx 2.264.$$

Figure 4.5, which displays the results of 10,000 trials in which a simulated roulette wheel was spun 200 times and the number of 0s recorded, shows that the distribution of the number of 0s in 200 spins is roughly normal. Accordingly, the Empirical Rule applies.

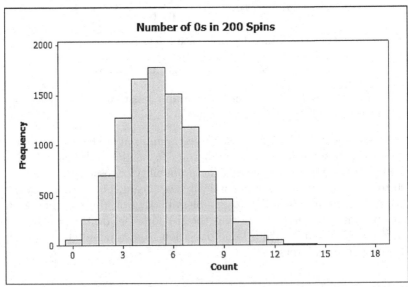

FIGURE 4.5: Results of 10,000 runs of 200 spins of a roulette wheel

Under the assumption that the wheel is truly random, the rule can be used to show that the probability that the number of 0s is more than 3 standard deviations above the mean is approximately .0015. Accordingly, if your wheel clocking shows any number appearing more than $\mu + 3\sigma \approx 12.054$ times in 200 spins, you may safely conclude that the roulette wheel favors that number, and you should bet in line with this discovery.

Sports Betting

Example 4.3.9. The Blue Chip Casino in Michigan City, Indiana, a location which does not allow sports wagering, nonetheless offered a free contest called "Pick the Pros" involving the 2012 National Football League season. Casino patrons were invited to pick the winner in each NFL game each week, and a top prize of $100,000 was to be awarded to anyone who successfully picked a least 12 winners correctly on at least 15 of the 17 weekends.

The probability of winning this jackpot is influenced by the following facts about the NFL season:

- In 2012, there were 32 teams in the league. Each team played 16 games, with one week off (the "bye" week) during the 17-week season.

- The bye weeks begin in the season's fourth week. Two teams have a bye in week 4, six have a bye in week 7, and four teams are idle every other week from the fifth through the eleventh. Accordingly, the number of games played each week ranges from 13 (one week) to 16 (eight weeks).

Assuming a constant 50% chance of picking each game correctly, the number of correct choices on a given weekend is a binomial random variable The probability of picking at least 12 winners in an N-game weekend is

$$\sum_{k=12}^{N} P(X=k) = \sum_{k=12}^{N} \left[{}_{N}C_{k} \cdot \left(\frac{1}{2}\right)^{k} \cdot \left(\frac{1}{2}\right)^{N-k} \right] = \sum_{k=12}^{N} \frac{{}_{N}C_{k}}{2^{N}}.$$

If $N = 16$, this probability is

$$\frac{2517}{65,536} \approx .0384$$

—not a very likely occurrence. For other values of N, we have the following probabilities, where X counts the number of games predicted correctly:

N	$P(X \geq 12)$
13	.0017
14	.0065
15	.0176
16	.0384

To win the top prize, you must pick at least 12 games correctly on at least 15 weekends. Your best chance for doing so is to be successful on all eight 16-game weekends, the one 15-game weekend, and six 14-game weekends, omitting a 14-game weekend and the one 13-game weekend. The probability of doing this is

$$p = (.0384)^8 \cdot (.0176) \cdot {}_7C_6(.0065)^6 \approx 4.393 \times 10^{-26}.$$

There are ${}_{17}C_{15} = 136$ different ways to choose 15 weeks to be successful from the 17 weeks in the season. To compute your total chance of winning, we would need to perform 136 calculations like the one above and add the results. (Many of those calculations would result in the same number, of course.) Since p as computed here is a maximum probability for success in any one of those 136 possibilities, an upper bound for your chance of winning the contest is given by

$$\hat{p} = 136p \approx 5.974 \times 10^{-24}.$$

The Blue Chip hedged its bets in its advertising by noting that "In the case of multiple winners, the \$100,000 prize will be split among the winners." In light of the long odds against one person winning, it seems highly unlikely that more than one will win, and so this clause stands very little chance of being activated.

4.4   EXERCISES

Answers begin on page 254.

4.1. *Sicherman dice* are a variant on ordinary six-sided dice. One die of the pair is numbered 1, 2, 2, 3, 3, 4; and the other is numbered 1, 3, 4, 5, 6, 8.

a. Find the probability of rolling a 7 with a pair of Sicherman dice.

b. Find the probability of rolling doubles with a pair of Sicherman dice.

c. Let X be the sum of the numbers rolled on a pair of Sicherman dice. Construct a probability distribution for X. How does this compare to the distribution for the sum of two standard dice seen in Example 4.1.5?

4.2. Consider a carnival game where you roll a d6 once and are paid, dollars, an amount equal to the number thrown. What would be the correct price to charge to make this a fair game?

4.3. A one-roll craps bet that the next roll will be an 11—called "yo-leven" at the tables—pays 15 to 1. Find the house advantage.

4.4. Another standard multiroll craps bet is the "Big Six" bet. This bet pays even money if a 6 is rolled before a 7. There is also a "Big Eight" bet—since the 6 and 8 are equally probable, the two bets are mathematically equivalent.

a. Find the house advantage for the Big Six bet.

b. At what rate would this bet have to pay off to make it a fair bet?

4.5. If the Big Six and Big Eight bets are not to your liking, a second option is to *place* those two numbers. Place bets get their name from the fact tha they are made for a player by a dice dealer; you place your chips in a neutral place on the layout and tell the dealer how you want them wagered. A place bet may be made on any point number and is simply a bet that that number will be rolled before a 7. This differs from a pass line bet in that the number you choose need only be rolled once, not once to establish the point and then a second time to win. Place bets pay off as follows: The 4 and 10 pay off at 9 to 5, the 5 and 9 at 7 to 5, and the 6 and 8 at 7 to 6. As always, fractions are rounded down, in favor of the house, so the wise gambler will make place bets in multiples of five or six, as appropriate.

Since a place bet on 6 or 8 pays 7 to 6 while the Big Six and Big Eight pay only even money, the place bet is clearly a better option for the craps player. Compute the HA of a $6 place bet (to avoid fractions) on 8 to determine how much better this bet is.

4.6. In European roulette, the HA on even-money bets (red/black, high/low, odd/even) is slightly lower than indicated in Example 4.2.7 when the *en prison* rule is used. Under this rule, if a player makes an even-money bet and the wheel turns up a 0, the bet is not resolved, but rather placed "in prison," to await the next spin. If the bet wins the next spin, the initial bet is returned to the player with no additional payoff; if not, the bet is collected. If the second spin is also a 0, the bet loses—*en prison* lasts for only one more spin. Find the house advantage for a $1 even-money bet with *en prison* in force.

One might reasonably wonder why a casino would offer the *en prison* option, which works to reduce the house edge. One possibility, as explained in [41], is that *en prison* is offered on the even-money bets, which are the bets most often involved in gambling systems for roulette (see Section 7.1). By decreasing the risk on even-money bets, the casino is subtly encouraging system players. Since betting systems do not work, this encouragement allows a casino to make up in volume what it gives back in a smaller HA.

4.7. The Michigan State Lottery also offers a Daily 4 game. Like the Daily 3, players pick a four-digit number for one of two drawings per day. Bets on this number may be made for 50¢ or $1.

- For a straight bet, the player wins if their number is exactly the one matched by the state. This pays off at 5000 for 1.

- Another option is the *wheel* bet, which, for $24, allows a straight bet on every permutation of a four-digit number with four different digits. In effect, this combines 24 separate tickets into one and is thus easier for both the gambler and the clerk who prints the ticket. If any one of the 24 permutations is drawn, the ticket pays $5000, or 5000 for 24.

- There are several varieties of boxed bets available.

    - A *24-way* boxed bet is made on a four-digit number with four different digits, such as 1729. If any combination of the four digits is chosen, the bet pays off at 208 for 1.

    - A *12-way* boxed bet covers a number which has two copies of one digit and two nonmatching digits, for example, 1146. If any of the 12 rearrangements of the number hits, the payoff is 416 for 1.

    - The *6-way* boxed bet is for a four-digit number with two pairs of identical digits, as in 6688. The payoff on a 6-way box is 833 fo

    - Finally, a *4-way* boxed bet can be made on a number where one of the two different digits is repeated three times, like in 3393. This payoff is 1250 for 1.

a. Compute the HA for the straight and wheel bets. How do they compare to the HA for the straight bets in the Daily 3 game (Example 4.2.21)?

b. How do the HAs for the various boxed bets compare to one another and to the HA for the straight bet on the same four-digit number?

4.8. Using the probabilities and prizes described in Example 2.4.18, and assuming a jackpot of \$1,000,000 (the minimum possible jackpot) compute the expected value of a \$1 Classic-47 lotto ticket sold by the Michigan State Lottery.

4.9. In Table 4.5, explain why switching the "Full house" and "4-card royal flush" lines would not change the strategy for a player following the table.

4.10. A parlay card offered by a consortium of seven casinos in south Nevada offers the payoffs indicated in Table 4.14.

TABLE 4.14: Alternate sports parlay card payoffs

Wins	Payoff
3 for 3	6 for 1
4 for 4	10 for 1
5 for 5	20 for 1
6 for 6	35 for 1
7 for 7	75 for 1
8 for 8	125 for 1
9 for 9	250 for 1
10 for 10	700 for 1

Let p be your probability of picking the winner of a single game.

a. What value of p will make the 3 for 3 bet a break-even proposition?

b. What value of p will make the 9 for 9 bet a break-even proposition? If you have that same probability p of picking one game correctly, do you have an advantage in playing the 10 for 10 wager? If so, how big is it?

4.11. Compute the house advantages for the bets other than on the $1 spot on the Big Six Wheel, using the Las Vegas payoffs stated in Example 4.2.24.

4.12. Example 3.2.19 described the Ninety Percenter punchboard. The punchboard has 800 spots, which are sold for 10¢ each. Players punching out any multiple of 10 between 10 and 650 receive a $1 prize, and the last spot on the board carries a $5 bonus.

Assuming that the advertised 90% return is accurate, calculate the expectation of the first bet on a fresh Ninety Percenter punchboard.

4.13. For the Ninety Percenter punchboard described in Exercise 4.12, calculate the expectation for the player punching out the second spot.

4.14. Recall that the probability of drawing a royal flush at video poker is approximately 1 in 40,000. Use this estimate to calculate the probability of getting exactly two royal flushes in 40,000 hands of video poker.

4.15. In Example 3.2.9, we considered the parlay card available in the sports books run by Station Casinos. Assuming a 50% chance of picking each game successfully, find the probability that you will pick exactly half of a ten-game parlay correctly.

4.16. Another way of setting two dice is designed to throw more 7s than the 1 in 6 fraction we expect from a random roll. This could be an advantage for the shooter on the come-out roll. In 400 rolls, how many 7s would be necessary to constitute a statistically significant excess?

4.17. In spider craps (see page 73), the payoff for a one-roll bet that the next roll will be a 2 is 55 to 1. Find the house advantage for this bet. How does the HA compare to the 13.89% HA for the corresponding wager at standard craps?

4.18. The spider craps equivalent of the craps field bet pays off if the next roll is 2–6 or 12–16, with a 2 to 1 payoff on a 16 and an even-money payoff on the other rolls. Find the house edge on this wager.

4.19. The one-roll *horn* bet in spider craps is derived from the standard craps menu of wagers. In standard craps, a horn bet wins if the next roll is a 2, 3, 11, or 12; in spider craps, the horn bet covers the numbers 2, 3, 15, and 16. Customarily, horn bets are made in multiples of 4 units, with equal bets on each number. At least three of these four bets will lose on each roll, but the fourth may win and cover the other losses. Bets on 2 or 16 pay off at 55 to 1; bets on 3 or 15 pay off at 28 to 1. Find the HA of a $4 horn bet at spider craps.

4.20. At what odds must the baccarat Tie bet pay off to reduce the HA on this wager to approximately 4%?

4.21. Suppose that a variation on baccarat provides that bets on Player or Banker both lose rather than push if the two hands tie. Find the HA for both bets with this rule in force.

Modified Casino Games

In the preceding chapters, we considered what might be called "standard" casino games. Game development specialists frequently propose new games or modifications to existing games or wagers, which are the focus of this chapter.

As of 2010, there are 506 table games approved for play in Nevada casinos [34, p. 179]. While the vast majority of new casino game ideas are unsuccessful, the potential to make a lot of money by selling the rights to a new popular game continues to attract game designers. For example, the inventors of Caribbean Stud Poker sold their idea for $30 million [34]. An important part of any new game proposal is a mathematical analysis of the bets and a careful calculation of the HA.

5.1 ROULETTE

Example 5.1.1. A new roulette betting option, introduced in 2008 and seen at the Orleans Casino in Las Vegas, is the "Colors" bet [7]. To make this bet, a gambler wagers on either Red or Black, and if that color then turns up on three consecutive spins, the payoff is 8 to 1. We would like to find the house advantage for a $1 Colors bet if an American (0 and 00) roulette wheel is used.

No matter which color is chosen, we have $P(\text{Win}) = p = \frac{18}{38}$ for a single spin. Since the spins are independent, the probability of winning the Colors bet is $p^3 = \left(\frac{18}{38}\right)^3 = \frac{729}{6859} \approx .1063$, and so we have

$$E = (8) \cdot \left(\frac{729}{6859}\right) + (-1)\left(\frac{6130}{6859}\right) = -\frac{298}{6859} \approx -\$.0434,$$

and so the house advantage is about 4.34%—somewhat less than the HA for any other bet on the board.

An analogous calculation for European roulette ($P(\text{Win}) = \frac{18}{37}$) wheels shows that the expected value of a $1 Colors bet is

$$E = (8) \cdot \left(\frac{5832}{50,653}\right) + (-1)\left(\frac{44,821}{50,653}\right) = \frac{1835}{50,653} \approx \$.0362,$$

about 3.6¢. Since this is positive, the player has a 3.6% advantage over casino—meaning that you won't see the Colors bet with this payoff on a European wheel. If you do, someone has made a serious calculation or judgment error. The bet is unfair in your favor, and you should settle in and prepare to win money.

Royal Roulette

Royal Roulette is a relatively new variation on roulette, developed in Australia, that uses familiar playing cards as spots on the wheel [57]. The wheel contains 50 pockets: one for each of the 48 cards from 2 through king, a single ace, and a joker. The layout (Figure 5.1) is similar to a traditional roulette layout and offers the betting options listed in Table 5.1.

Joker	♡ ♣ 2 ♠ ◇	♡ ♣ 4 ♠ ◇	♡ ♣ 6 ♠ ◇	♡ ♣ 8 ♠ ◇	♡ ♣ 10 ♠ ◇	♡ ♣ Q ♠ ◇	3:1 ♡
							3:1 ♣
A	♡ ♣ 3 ♠ ◇	♡ ♣ 5 ♠ ◇	♡ ♣ 7 ♠ ◇	♡ ♣ 9 ♠ ◇	♡ ♣ J ♠ ◇	♡ ♣ K ♠ ◇	3:1 ♠
							3:1 ◇
1ST 16		2ND 16		3RD 16			
2-7	EVEN	Red	Black	ODD	8-K		

FIGURE 5.1: Royal Roulette layout

TABLE 5.1: Royal Roulette betting options

Name	Payoff	Description
Straight Up	47 to 1	A bet on any single card including Ace/Joker
Split	23 to 1	A bet covering 2 cards
Court	15 to 1	A bet covering 3 cards (Ace3♠3♡ or Joker2♠2♡)
Street/Corner	11 to 1	A bet covering 4 cards
Six Line	7 to 1	A bet covering 6 cards (Ace, Joker, 2♡, 2♠, 3♡, 3♠)
Eight Line	5 to 1	A bet covering 8 cards
Suit/Column	3 to 1	A bet covering 12 cards (One suit or one column)
Sixteen Set	2 to 1	A bet covering 16 cards
Even Chances	1 to 1	Red, Black, Odd, Even, 2-7, 8-K (24 cards each)

Example 5.1.2. How does Royal Roulette compare to the standard game on which it is based? A quick examination of the payoff structure shows that the payoff on a winning bet on n cards is $\dfrac{48-n}{n}$ to 1. There is no exception corresponding to the basket bet in American roulette. Royal Roulette's versions

of the basket bet, by which we mean unusual bets derived from the layout, are the Court and Six Line bets. Unlike in standard American roulette, the Royal Roulette analogs pay in line with other bets. We can compute the expected value of any Royal Roulette bet with one simple equation:

$$E = \left(\frac{48-n}{n}\right) \cdot \left(\frac{n}{50}\right) + (-1) \cdot \left(\frac{50-n}{50}\right) = \frac{48-n-50+n}{50} = -\frac{2}{50} = -\$$$

—for a house edge of 4%, a result which is independent of n and more favorable to the player than American roulette.

A second version of Royal Roulette removes the joker from the wheel and the layout—we might think of this, informally, as European Royal Roulette. The Six Line bet is eliminated, though the Court bet remains and must involve the Ace. Though the number of pockets on the wheel is now only 49, the payoffs are the same as in the jokered game, and the expectation can again be calculated for all bets at once:

$$E = \left(\frac{48-n}{n}\right) \cdot \left(\frac{n}{49}\right) + (-1) \cdot \left(\frac{49-n}{49}\right) = \frac{48-n-49+n}{49} = -\frac{1}{49} \approx -\$.0204$$

which yields a game with a lower casino edge, 2.04%, than European roulette.

Diamond Roulette

In 2008, a roulette variant called *Diamond Roulette* debuted in Atlantic City, New Jersey. In Diamond Roulette, the two-color red and black layout is replaced by one with six colors: black, blue, green, red, purple, and yellow. Each color corresponds to six numbers on the wheel: for example, the numbers 8, 12, 18, 19, 29, and 31 are purple on a double-zero Diamond Roulette wheel [49]. 0 and (if present) 00 remain green but are not counted among the green numbers for the purpose of color betting.

In order to accommodate bets on the six colors, the traditional roulette layout is modified to remove the red, black, and dozens bets on a standard layout. The other standard roulette bets (Table 2.2) are all available and pay off as usual to produce a 5.26% HA (7.89% for the basket bet).

The six bets on the individual colors, being mathematically equivalent to the double street bet, also pay off at 5 to 1 if a number of the selected color is spun and carry the same 5.26% house advantage on a double-zero wheel or 2.70% on a single-zero wheel.

Example 5.1.3. Perhaps in an effort to attract gamblers who understand that the 0 and 00 are the source of the casino's advantage, a new bet offered at double-zero Diamond Roulette is the *Double Diamond* or *eight numbers color* wager, which allows a bettor to wager on a single color and which pays off if a number of that color or either of the two zeros comes up on the wheel

[50]. The eight-number bet pays off at 3 to 1, and so the expected value of a $1 wager is

$$E = (3) \cdot \left(\frac{8}{38} \right) + (-1) \cdot \left(\frac{30}{38} \right) = -\frac{6}{38} \approx -\$.1579,$$

giving the casino a 15.79% advantage, triple that of almost every other bet—which makes this a bet to be avoided.

Riverboat Roulette

2012 saw the debut of *Riverboat Roulette*, a game variation that also expanded the roulette spectrum beyond red and black, but with a new twist for roulette. In addition to retaining their red or black colors for the purposes of standard color wagering, 32 of the 36 numbers from 1 to 36 are also tagged with a second color, which appears in the pocket instead of behind the number on the wheel:

Color	Numbers
Teal	5, 17, 32
Orange	4, 16, 21, 23, 33
Pink	8, 12, 19, 29
Purple	7, 11, 20, 30
Blue	3, 15, 22, 24, 34
Yellow	6, 18, 31
White	1, 2, 9, 10, 13, 14, 27, 28

The numbers 25 and 36 remain red, and 26 and 35 remain black, without a second color.

By reference to Figure 5.2, one can see that each new color group except for white is comprised of a block of adjacent numbers on the wheel, and so these color groups are called *neighborhoods*. Each set of two numbers immediately adjacent to one of the zeroes, on either side, is colored white.

It is a point of interest that coloring adjacent numbers the same color might make Riverboat Roulette susceptible to wheel clocking, as described in Section 4.3.

Riverboat Roulette allows for all of the bets available on a standard roulette wheel—the red, black, and

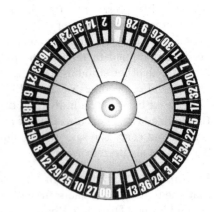

FIGURE 5.2: American roulette wheel

dozens bets are not removed as is the case in Diamond Roulette—and adds a new type of wager on six of the seven new neighborhoods. Bets on the white neighborhood are one-spin bets, like bets on red or black; see Exercise 5.12 for details. In ordinary roulette and the variants we have seen so far, each spin of the wheel resolves all bets, and the layout is cleared before the next spin. In Riverboat Roulette, bets on the neighborhoods other than white function like place bets (Exercise 4.5) in craps. A bet on one of the other six colors is a bet that a number in that neighborhood will be spun before a white number, and if a number of any other color—including red, black, or green—is spun, the result is a push, and the bet remains active. If a white number is spun, this is called a "White Out," and all neighborhood bets lose; this is roughly analogous to sevening out at craps. Neighborhood bets may be "taken down" (withdrawn) or increased following a push.

Example 5.1.4. What is the house advantage of these new neighborhood bets?

For each color, we can focus on the wheel numbers that resolve the bet: the eight white numbers and the three, four, or five numbers in the neighborhood. Denote this latter number by x. The neighborhood bets have the following values for x and corresponding payoffs:

Color	x	Payoff
Teal & Yellow	3	2 to 1
Pink & Purple	4	8 to 5
Orange & Blue	5	7 to 5

For the teal and yellow neighborhoods, there are 11 numbers that resolve the bet. The expectation on a $1 bet is

$$E = (2) \cdot \left(\frac{3}{11}\right) + (-1) \cdot \left(\frac{8}{11}\right) = -\frac{2}{11} \approx -.1818,$$

and the HA is 18.18%.

In the remaining two cases, we assume a $5 wager to eliminate fractions in the payoff. For pink and purple neighborhoods, 12 numbers resolve the bet.

$$E = (8) \cdot \left(\frac{4}{12}\right) + (-5) \cdot \left(\frac{8}{12}\right) = -\frac{8}{12} \approx -.6667.$$

Dividing by the $5 wager gives an HA of 13.33%.

Orange and blue wagers involve 13 numbers. The expectation on a $5 bet is

$$E = (7) \cdot \left(\frac{5}{13}\right) + (-5) \cdot \left(\frac{8}{13}\right) = -\frac{5}{13} \approx -.3846,$$

so the HA is 7.69%.

While these bets may increase gamblers' interest in roulette through the different betting options and the novelty of multi-spin wagers, it seems unlikely that wagers with such a high house edge will attract much action. The option to take down a neighborhood bet after a push may make these bets slightly less unattractive.

It is worth noting that these percentages are the house edge *per resolved bet*. An alternate interpretation of the phrase "house advantage" is the HA *per spin of the wheel*. This calculation considers all 38 numbers on the wheel, including those for which the wager is a push and the payoff is 0. Denote the expectation per spin by E', to distinguish it from the expected values already computed. For the teal and yellow neighborhoods, the expectation per spin is

$$E' = (2) \cdot \left(\frac{3}{38}\right) + (-1) \cdot \left(\frac{8}{38}\right) + (0) \cdot \left(\frac{27}{38}\right) = -\frac{2}{38} \approx -\$.0526,$$

and so the recalculated HA is the 5.26% we're accustomed to in roulette. It can be shown that a bet on the teal or yellow neighborhoods will stay on the layout for an average of $3\frac{5}{11}$ spins before being resolved; multiplying this number by 5.26% returns the 18.18% HA found originally.

Similar calculations will show that for the pink and purple neighborhoods, the expected return per spin on a \$5 bet is $E' = -\$.2105$, for a new HA of 4.21%—better than 5.26%—and for a \$5 bet on orange or blue, $E' = -\$.1315$, so the HA there is 2.63%, half the edge of a standard roulette wager.

Double Action Roulette

Also in 2012, the M Resort in Henderson, Nevada introduced *Double Action* roulette. A Double Action wheel contains two identical rings of 37 or 38 numbers (in either European or American roulette configurations). The two wheels rotate independently in opposite directions, and when the ball falls into a pocket, it identifies two winning numbers, one on each wheel.

Two separate betting layouts allow standard roulette bets (with an HA of 2.70% or 5.26%) on either wheel separately, and a collection of new bets that apply to the numbers on both wheels. These bets mimic the standard roulette bets; one may bet, for example, on whether or not both numbers will be red, or even, or low.

Example 5.1.5. One new bet is a separate single-number *side bet* (a bet unrelated to the main play of a casino game) that pays off at 1200 to 1 when the chosen number is hit on both wheels in a single spin. The expectation on a \$1 bet on this proposition, at an American Double Action wheel, is

$$E = (1200) \cdot \left(\frac{1}{38}\right)^2 + (-1) \cdot \left[1 - \left(\frac{1}{38}\right)^2\right] = -\frac{243}{1444} \approx -\$.1683,$$

a 16.83% house edge. The edge for this bet on a European Double Action wheel is not much better: 12.3%. In the first month of Double Action Roulette's test run at M, there were 18 of these jackpots paid [65].

Double Action is a direct descendant of *Double Roulette*, a game introduced at Monte Carlo in 1936, which used two concentric wheels and two roulette balls [12]. The high house advantage doubtlessly contributed to the game's short lifespan in Monaco.

5.2 DICE GAMES

Crapless Craps

Crapless craps is a variation that was developed by Bob Stupak, owner of the Vegas World Casino (now the Stratosphere) in Las Vegas. If 2, 3, 11, or 12 is rolled on the come-out roll, they neither lose nor win, but become points like any other point, and the shooter must roll that number again before rolling a 7 to win. A 7 on the come-out roll remains an automatic winner. To make up for this change in rules, the Don't Pass bet is not offered at crapless craps—you cannot bet against the shooter.

Example 5.2.1. On the face of it, this looks like a good deal for dice players: four automatic losing come-out rolls (1-1, 1-2, 2-1, and 6-6) are now points with a chance to win, while only two automatic wins (5-6 and 6-5) have been converted to possible losers. The mathematics tells a different story, though. In addition to the figures in Example 3.2.6, we have the following new probabilities:

Point	P(Point rolled)	P(Point wins)	P(Win on this point)
2 or 12	1/36	1/7	1/252
3 or 11	2/36	2/8	1/72

So the probability of winning after a come-out roll of 2 or 12 has risen from 0 to $\frac{1}{7}$, and the probability of winning after a 3 has risen from 0 to However, the probability of winning after an 11 has fallen from 1 to $\frac{1}{4}$. The player has given up far more than he has gained. Adding everything up reveals that the probability of winning a Pass line bet at crapless craps is .4730, and thus that the house advantage has risen to approximately 5.39%—making this game about as advantageous for the casino as American roulette.

Barbooth

Bob Stupak was known for offering unusual games at Vegas World. In the game of *barbooth*, he took a street game called *barbudi*, a Canadian game also favored in large northern U.S. cities, and revamped it for casino play. Barbooth is unusual in that it is a fair game, with neither side having an advantage [59].

In barbooth, two players alternate turns at throwing two dice. A shooter, and everyone betting on him or her, wins when tossing 3-3, 5-5, 6-6, or 5-6; the probability of this is $\frac{5}{36}$. The shooter and any backers lose when 1-1, 1-2, 2-2, or 4-4 is rolled; an event that also has probability $\frac{5}{36}$—making the game even. All other rolls result in the dice being passed to the other player. A winning player in one round shoots first in the next round. In the casino, these two players are designated "Player" and "Bank," as is also the case in baccarat.

Example 5.2.2. How many rolls, on average, are necessary for a round of barbooth to be resolved?

Define $p = \frac{10}{36}$, the probability of the game ending on any one turn, and let $q = 1 - p = \frac{26}{36}$. The probability of the game ending on the first roll is p. The probability of the game lasting two rolls is $q \cdot p$, where the first roll neither wins nor loses and the second roll finishes the game one way or the other. In general, a game has probability $q^{k-1} \cdot p$ of lasting exactly k rolls.

The expected number of rolls is then

$$E = \sum_{k=1}^{\infty} k \cdot q^{k-1} \cdot p$$

$$= \frac{p}{q} \cdot \sum_{k=1}^{\infty} k \cdot q^k$$

$$= \frac{p}{q} \cdot \left(\frac{q}{(q-1)^2} \right),$$

using the computer algebra system Mathematica for the summation. Continuing, we have

$$E = \frac{p}{p^2}, \text{ since } q - 1 = -p$$

$$= \frac{1}{p}$$

$$= \frac{36}{10},$$

or 3.6 rolls.

With no edge on either side, where does the casino make its money? Basic bets at Barbooth pay off at less than even money. As in baccarat, a 5% commission is charged on all bets and collected on losing bets. This is accomplished by paying off bets at 20 to 21 odds: players risk $21 but are paid only $20 if they win. At the same time, casino barbooth offers side bets, similar to the wagers found on a craps table, that also charge the 5% commission. The craps bet "Any 7" is often available, as are separate bets on the eight decisive rolls.

Free Odds

Free odds bets in craps, though commonly available, are a casino rarity—a bet with zero house advantage. Once a point is established on the come-out roll, players who have placed a pass line bet have the opportunity to "back it up" with an additional *odds* bet. These odds bets are limited to a certain multiple of the original bet, often "3X/4X/5X," meaning three times the original bet if the point is 4 or 10, four times the original bet if the point is 5 or 9, and five times the original bet if the point is 6 or 8. The exact multiple allowed is decided by the individual casino; in 2012, as part of a new focus on player-friendly gambling in their advertising, the Riviera Casino in Las Vegas offered 1000X odds bets on any point in an effort to entice craps players to their casino.

Putting the "free" in free odds is this: These bets are fair bets, without any house advantage. An odds bet on a point of 4 or 10 pays 2 to 1, since the probability of making a 4 or 10 is $\frac{3}{9} = \frac{1}{3}$. Similarly, an odds bet on 5 or 9 pays off at 3 to 2, and odds bets on 6 or 8 are paid at 6 to 5. It is wo noting that, here as elsewhere in a casino, fractions of a dollar that cannot be accommodated by common chip values—casino chips with values less than 50¢ are unusual; some casinos use half-dollar coins for this denomination—are rounded *down*, in favor of the casino. To gain the full effect of free odds, it is important to bet in such a way that no rounding is necessary. The easiest way to do this is to remember to place odds bets in multiples of $10.

Example 5.2.3. From a gambler's perspective, then, pass line bets should be made to get as little as possible on the front line and as much as possible in an odds bet. If the casino has a $5 minimum on the pass line and you wish to bet $25 on each decision, the best possible play is to make a $5 line bet and, once the point is established, back it up with a $20 odds bet, provided that the multipliers allow that. Recall that the HA on a pass line bet is 1.41%. Using free odds in this manner, and assuming that the casino offers at least 4X odds on every point (for simplicity), the expected return on the bet is

$$E = 5 \cdot \left(\frac{8}{36}\right) - 5 \cdot \left(\frac{4}{36}\right) + 45 \cdot \left(\frac{2}{36}\right) + 35 \cdot \left(\frac{4}{45}\right) + 29 \cdot \left(\frac{50}{396}\right) - 25 \cdot \left(\frac{196}{495}\right) = -$$

which corresponds to an HA of merely .28%. The first two terms in the middle of this equation correspond to those outcomes when the bet is resolved on the come-out roll and there is no opportunity to place an odds bet; the remaining terms represent the different ways that the bet with full odds can play out.

Why would a casino offer odds bets? One reason is to encourage activity at the craps tables, of course. Another is to encourage larger bets and take advantage of volatility. We note that the only time that an odds bet can be made is *after* a point has been established, which is when the casino has the advantage over a pass line bettor.

Free odds are also available on Don't Pass bets; since the Don't Pass bettor has the advantage once the point is established, these pay off at less than even money: 1 to 2 if the point is 4 or 10, 2 to 3 on 5 or 9, and 5 to 6 on 6 or 8. Moreover, odds bets backing up a Don't Pass bet may be taken down by the bettor at any time before they're resolved—the casino is pleased to let gamblers back out of a bet where they have a greater chance of winning than of losing.

Example 5.2.4. A pass line player taking full advantage of the 1000X bets at a \$5 minimum table at the Riviera would be risking \$5005 on each established point. Some of these bets will be lost to the casino, even though the expectation remains

$$E = 5 \cdot \left(\frac{8}{36} \right) - 5 \cdot \left(\frac{4}{36} \right) + 10,005 \cdot \left(\frac{2}{36} \right) + 7505 \cdot \left(\frac{4}{45} \right)$$
$$+ 6005 \cdot \left(\frac{50}{396} \right) - 5005 \cdot \left(\frac{196}{495} \right)$$
$$= -\frac{7}{99}.$$

Neither the odds of winning nor the expected value has changed, since the free odds bet is paid at true odds. Only the magnitude of the bets involved is different. The HA has dropped to a mere .0014%, but every losing odds bet nets the casino \$5005, and it is quite likely that many dice players will not be able to withstand too many consecutive losses at these stakes.

The mean and standard deviation can be used to assess the volatility of this betting option. Consider the case where the point is 6 (or 8). For the wagers in which a full odds bet is made—that is, where the bet is not resolved on the come-out roll—we have the following probability distribution for the winnings X:

x	6005	−5005
$P(X = x)$	$\frac{5}{11}$	$\frac{6}{11}$

This distribution gives $\mu = -\$.45$ and $\sigma = \$5482.20$. Note the extremely high value of σ relative to μ—this indicates a bet with very high variability and thus the potential for wild swings back and forth. Since the casino has far more money than any player, it is much better equipped to withstand a run of bad luck than the gambler is.

Craps Side Bets

It may be said that *every* bet on a craps table except for the Pass and Don't Pass wagers is a side bet, in the sense that they are bets that are not related to the main play of the game. The bets described in this section are relatively new bets that do not have dedicated spaces on the standard craps layout (Figure 3.1).

Example 5.2.5. The *Fire Bet* is a craps side bet developed by Las Vegas casino supervisor Perry Staci, whose name suggests its purpose: to cash in on a shooter having a "hot" hand and making many points before sevening out [6]. A bet on the Fire Bet pays off if the shooter makes a certain number of *different* points, at least three or four depending on the paytable, bef tossing a 7. While the paytable for the Fire Bet varies from casino to casino, one version is this:

Points Made	Payoff
4	10 to 1
5	200 to 1
6	2000 to 1

The points made must be different: if the shooter makes the points 4, 9, 4, and 6 before sevening out, this is only three different points for the purpose of the Fire Bet. A roll of 7 on a come-out roll does not interrupt a string of points. Mathematically, the Fire Bet is somewhat complicated to analyze, due to the different probabilities for establishing and making the six possible points. A good source for the computations required is [70]; there we find the following:

Points Made	Probability
0	.5939
1	.2608
2	.1013
3	.0334
4	.0088
5	.0016
6	.000162

With the paytable above, the probability of winning anything with the Fire Bet is a mere .0106, or 1.06%. The corresponding expectation is

$$E = (-1) \cdot .9894 + (10) \cdot .0088 + (200) \cdot .0016 + (2000) \cdot .000162 = -.2486$$

giving an HA of 24.9%.

In part, no doubt, because of the high casino advantage, the Fire Bet has gained a place on the craps table in many casinos. Despite the high HA, gamblers seem to be drawn to it; they are perhaps attracted in part by the large payoff for a small bet. The maximum Fire Bet wager at most casinos offering this wager is $5.

Example 5.2.6. In 1997, Donald Catlin and Leonard Frome proposed a new craps wager called the *Hard Hardway* bet. This is a simple wager that each of the six even sums, 2 through 12, will be rolled "the hard way" (as doubles) twice before a 7 is rolled. As one might expect, this is an extremely unlikely event. Considerable calculation (see [8]) leads to a probability of winning of

$$p = \frac{9,198,254,528,424}{1,458,015,678,282,240,000},$$

which is approximately 1/158,510.

On one hand, this seems so unlikely that no one would ever make this bet. On the other hand, the probability of winning a Powerball lottery jackpot is considerably less, and people still buy lottery tickets, so it's possible that, as with the Fire Bet, a large potential payoff might attract bettors in spite of the unfavorable odds. Suppose that a casino offers the Hard Hardway bet with a payoff of 150,000 to 1 for convenience—theirs. The expected value for a Hard Hardway bet with this payoff is

$$E = (150000) \cdot p + (-1) \cdot (1 - p) \approx -\$.054,$$

where p is the value above, and so the house advantage is about 5.4%. W this is more unfavorable to a player than a Pass or Don't Pass bet, the house advantage here is considerably less than that of several craps bets that are currently offered.

Double Dice

Double Dice was introduced at the Majestic Star Casino in Gary, Indiana, in 2010. The goal in Double Dice is to roll a number as close as possible to 12 without going over, on either two or three standard dice. The game is played as follows:

- Two players each roll 2d6, hoping for the highest possible total. A 12 on the first roll is considered a perfect score. If both players roll a 12, then the game is a tie. If only one player rolls a 12, then that player wins.

- If there are no 12s, the player with the lower score gets to roll an additional die to improve his or her score. If both players are tied below 12 after the initial roll, both players get to roll a third die.

- At this point, the score closest to 12 without going over is the winner. If the two players have the same score, the game is a tie unless they both have 11, which is a loss for both players.

Players rolling the dice must make a Primary Bet that they will win, which pays off at even money. If their roll exceeds 12, the Primary Bet loses. A range of bets is available to them and to others wishing to place a bet:

Wager	Description	Payoff
Double Out	Both players exceed 12	80 to 1
Perfect Dice	Both players roll perfect 12s	50 to 1
	One player rolls a perfect 12	15 to 1
Double Dice	Both players roll identical pairs, 1-5	10 to 1
	One player rolls a pair	2 to 1

With regard to the Perfect Dice bet, a 12 at Double Dice is similar to a 21 in blackjack: this bet only pays off if the players roll 12s on the first roll. A 12 on three dice is a 12, not a "Perfect 12," and thus this bet does not pay off.

Example 5.2.7. The expectation of a $1 Perfect Dice bet is

$$E = (50) \cdot \left(\frac{1}{36}\right)^2 + (15) \cdot \left(2 \cdot \frac{1}{36} \cdot \frac{35}{36}\right) + (-1) \cdot \left(\frac{1225}{1296}\right) = -\frac{125}{1296} \approx -\$.0965$$

so the casino holds a 9.65% edge on this bet.

The Primary Bet ought to be exactly even. The casino draws its edge from the handling of "over 12" rolls and the rule that ties on 11 lose. Taken together, these rules give the house an advantage of 2.784% on this bet.

Chuck-A-Luck and Sic Bo

We noted in Section 4.2 that chuck-a-luck's high HA of 7.9% might be responsible for its disappearance from casinos. Assuming the truth of this hypothesis, game designers might reasonably set out to modify the game and reduce the house advantage in order to attract players. There is a delicate balancing act at work here, in that changing the game so much that the gambler has an advantage will result in a game that casinos will not offer.

Example 5.2.8. One possibility would be to introduce a "bonus die" that pays out twice what the other dice do. Suppose that we replace the three identical dice in the standard cage with two white dice and one purple die, with the purple die paying off at 2 to 1 if it shows the player's number.

In this game, the probability of losing $1 remains

$$P(-1) = \left(\frac{5}{6}\right)^3 = \frac{125}{216}.$$

The remaining probabilities change, depending on whether or not the chosen number appears on the purple bonus die. For a $1 win, the number cannot appear on the bonus die, so

$$P(1) = \left[2 \cdot \frac{1}{6} \cdot \frac{5}{6}\right] \cdot \frac{5}{6} = \frac{50}{256},$$

where the term in brackets computes the probability of rolling the chosen number on exactly one of the two white dice, and the remaining 5/6 factor accounts for not rolling the number on the bonus die.

For \$2, there are two cases to consider: either the number appears on both white dice or only on the purple die. These cases may be combined using the First Addition Rule, so

$$P(2) = \left[\frac{1}{6} \cdot \frac{1}{6} \cdot \frac{5}{6}\right] + \left[\frac{5}{6} \cdot \frac{5}{6} \cdot \frac{1}{6}\right] = \frac{5+25}{216} = \frac{30}{216}.$$

A \$3 win occurs when one white die and the purple die show your number, and the third die does not.

$$P(3) = 2\left(\cdot\frac{1}{6}\right)^2 \cdot \frac{5}{6} = \frac{10}{216}.$$

Finally, you win \$4 when all three dice show your number. This is identical to the case of winning \$3 at standard chuck-a-luck and so has probability —

The resulting probability distribution for the random variable X that counts the winnings on a \$1 bet on a single number is

x	−1	1	2	3	4
$P(X = x)$	$\frac{125}{216}$	$\frac{50}{216}$	$\frac{30}{216}$	$\frac{10}{216}$	$\frac{1}{216}$

We note that the probability of losing \$1 has not changed. The probability of winning \$1 has fallen by $\frac{25}{216}$, with that fraction being distributed to higher payoffs, so the expected value of a bet will increase. The expected return can be calculated to be \$.088—so we have overcorrected and created a game with a positive player expectation of 8.8%.

Example 5.2.9. There are, of course, other ways to modify chuck-a-luck. If we just swap out the standard dice for eight-sided dice, we can easily see that the HA for three identical dice will increase, since with more sides, the probability of winning will decrease while the payoffs will still be \$1 to What if we incorporated the bonus die idea from Example 5.2.8, but with d8s?

With one bonus die, paying \$2, and two regular dice, paying \$1 each, it turns out that there are too many losing combinations. The probability distribution is

x	−1	1	2	3	4
$P(X = x)$	$\frac{343}{512}$	$\frac{98}{512}$	$\frac{56}{512}$	$\frac{14}{512}$	$\frac{1}{512}$

and the expected value comes out to -\$.16992, about a 17% HA—worse than regular chuck-a-luck.

So we try again, switching this time to one even money die and two that pay at 2 to 1. Things look better:

x	-1	1	2	3	4	5
$P(X = x)$	$\frac{343}{512}$	$\frac{49}{512}$	$\frac{98}{512}$	$\frac{14}{512}$	$\frac{7}{216}$	$\frac{1}{512}$

The expectation for this game is $-\frac{23}{512} \approx -\$.045$. With a 4.5% HA, we have found a modification of chuck-a-luck that—from a purely mathematical standpoint—might be competitive against roulette and most of the center bets in craps. The success or failure of a new game, of course, is a function of many other variables, most of which are not purely mathematical.

Sic bo is a three-die game that expands on chuck-a-luck. The chuck-a-luck option of betting on a single number from 1 to 6 remains, and it is joined by a host of additional betting options. For example, one may bet on various propositions involving the sum of the three dice, the occurrence of a specific pair of numbers, or on whether or not the dice will all show the same number. Figure 5.3 shows a betting layout for sic bo. At some casinos, including the Mohegan Sun in Montville, Connecticut, the layout is electrified, and the sic bo dealer need only press buttons corresponding to the numbers on the three dice to light up all of the winning bets.

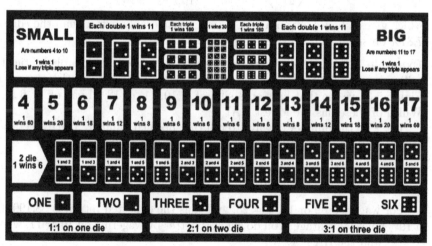

FIGURE 5.3: Sic bo betting layout [76]

Example 5.2.10. A new option in sic bo is a bet on a specified pair of different numbers, which pays off at 6 to 1 if both numbers appear among the three dice. Among the 216 ways for three dice to land, how many show a given pair of numbers, and what is the expectation of this bet?

The two numbers selected are immaterial, so we shall focus on the 1-2 bet. Designating the dice by Red, Green, and Blue, we have the following winning combinations:

Red	Green	Blue
1	2	Any
2	1	Any
1	Any	2
2	Any	1
Any	1	2
Any	2	1

Since each of the "Any" slots may be filled by six numbers, it appears that there are 36 winning combinations, which would give the player a 16.67% advantage. It follows, therefore, that we have missed something with this model. Certain winning combinations where "Any" is replaced by a 1 or 2 have been counted twice in this accounting; these are 1-2-1, 1-2-2, 2-1-1, 2-1-2, 1-1-2, and 2-2-1. Removing the duplicate occurrences of these six combinations from the list of 36 leaves 30 winning rolls and 186 losing rolls for the 1-2 bet; the resulting expectation on a $1 bet is

$$E = (6) \cdot \left(\frac{30}{216}\right) + (-1) \cdot \left(\frac{186}{216}\right) = -\frac{6}{216} \approx -\$.0278,$$

which gives a house advantage of 2.78%, considerably better than the edge of 7.87% on single-number chuck-a-luck bets.

Example 5.2.11. Galileo's work in determining the correct probabilities fo rolling sums of 9-12 on three dice is essential to sic bo bets on those sums. In Figure 5.3, we see that bets on a sum of 9, 10, 11, or 12 all pay off at 6 to 1. Which of these has the lowest HA?

Let x be the number of ways to roll the specified number. The expectation of a $1 wager is then

$$E(x) = (6) \cdot \left(\frac{x}{216}\right) + (-1) \cdot \left(\frac{216 - x}{216}\right) = \frac{7x}{216} - 1.$$

For wagers on 9 or 12, $x = 25$, and thus the expected return is $-\$.190$, for a 19% house advantage. You do slightly better by wagering on 10 or 11, where $x = 27$ and the HA is 12.5%.

Example 5.2.12. (An unintentional modification) The most exciting day in the history of sic bo was October 26, 1994 [100]. On that day, the Grand Casino in Biloxi, Mississippi was found to be unwittingly offering an 80 to 1 payoff for winning bets that the sum of the dice would be 4 or 17. Since there are three ways to roll a total of 4 on three dice (1-1-2, 1-2-1, and 2-1-1), the expected value of a $1 bet on 4 was

$$E = 80 \cdot \left(\frac{3}{216}\right) + (-1) \cdot \left(\frac{213}{216}\right) = \frac{27}{216} = \$.125$$

—*a 12.5% player advantage.* The same player advantage held for a bet on 17.

News of this unusual opportunity spread through the gambling community, largely via fax—this was in the infancy of the World Wide Web—and on the 26th, gamblers from across America converged on Biloxi, including one Minnesota man who drove for 17 hours to reach the casino. They played the game for hours, crowding the table so deeply that it was difficult to find a place. After the game shut down for the night, casino officials estimated that the casino's losses were in the neighborhood of $180,000. Casino management closed the game down the next day, and when it reopened, the payout on 4 and 17 had been reduced to the standard 60 to 1, resulting in a new expectation of

$$E = 60 \cdot \left(\frac{3}{216}\right) + (-1) \cdot \left(\frac{213}{216}\right) = -\frac{33}{216} = -\$.1528,$$

or an HA of about 15.3%.

5.3 CARD GAMES

Card Craps

In California and Oklahoma, state law holds that dice cannot be used as the sole device determining the outcome of a game of chance. In those two states, casinos may legally offer a game called *card craps*, a game similar to craps in which a special deck of cards is used to mimic a pair of dice.

In Oklahoma, a game developed by the Quapaw tribe uses two rows of six cards, each running from ace through 6 and randomly dealt facedown to the table. Players call out two numbers, which determine which cards from each row will be turned over and added to simulate a roll of dice. When the two rows of cards are shuffled and replaced after each simulated roll, this game is mathematically identical to standard craps [26].

A California version of card craps differs in both the number of cards in use and the probabilities of the various outcomes: either 264 or 324 cards are used, evenly distributed among the numbers 1 to 6. Two cards are dealt to simulate a roll of two dice. The deck is not dealt down very far without replacement; cards are immediately reinserted into a continuous shuffling machine that replaces them into the deck, but there may be a small time lag during which the numbers from 1 to 6 are not exactly equally likely. Specifically, cards that have just been dealt—perhaps as many as 12—are unavailable because of how the machine operates: these cards are held in a buffer and cannot be immediately redrawn so that the machine can dispense cards as needed without an undesirable time lag.

Example 5.3.1. One difference that is immediately apparent and that does not depend on any buffered reserve of cards is that the probability of "rolling" doubles is slightly less than $\frac{1}{6}$, because once the first card is drawn, the six numbers are no longer equally likely for the draw of the second card. Specifically, if a 264-card deck (44 of each rank) is used, the probability p of doubles is the probability that the second card matches the first in rank:

$$p = \frac{43}{263} \approx .1635 < \frac{1}{6} \approx .1667,$$

and with 324 cards (54 of each rank),

$$p = \frac{53}{323} \approx .1641 < \frac{1}{6}.$$

On a more practical level, the use of a continuous shuffling machine with a delay may make the Don't Pass bet more lucrative, with a positive expected value that favors the player who counts cards and adjusts odds bets with the count, taking them down when the count is bad and restoring them when the count is favorable [30]. Remember that odds bets on the Don't Pass line may be removed at any time before being resolved.

Example 5.3.2. Suppose, for example, that the point is 4. If a lot of 1s and 2s are dealt out of the deck and held in the buffer, they're unavailable to be dealt immediately, and so the cards will tend to generate higher totals: 7 will be even more favored relative to 4 than it is normally.

EZ Baccarat

The baccarat requirement that winning bettors on the Banker hand pay a 5% commission is a source of some confusion to players and dealers alike. *Baccarat* is an effort to eliminate this confusion by a rule change that changes the edge on a Banker bet from the bettor to the casino: If the Banker hand is a three-card total of 7, then a Banker bet pushes rather than wins against a lower Player hand. A Player wager still loses against a three-card 7 if the Player hand is 0-6.

The odds of a three-card 7 are approximately 43.4 to 1 [14]. This change in the rules, turning a winning hand into a push, is enough to give the house a 1.02% advantage when winning Banker bets are paid at 1 to 1 [37]. As a result, the need to pay commissions is eliminated, as is (more importantly) the time spent collecting them. With an increase in game pace, the casino can deal more hands per hour and so increase its long-term gain.

Example 5.3.3. Often appearing together with EZ Baccarat is the *Dragon 7* side bet, which is a simple bet that the Banker hand will win with a three-card 7 and which pays 40 to 1. The expectation of this wager is

$$E = (40) \cdot \frac{1}{44.4} + (-1) \cdot \left(1 - \frac{1}{44.4}\right) = -\frac{3.4}{44.4} \approx .0766,$$

giving the house a 7.66% advantage. In light of the fact that the HA on the main game is less than 1.25% on either Player or Banker, this is a bet to be avoided, though it is better than the Tie bet.

Casino War

Casino War is a gambler's version of the children's card game War. Playe and dealer are each dealt one card from a six-deck shoe (the number of decks may vary), and the high card wins. Aces are always high. At this point, the game is even—both sides have an equal chance of drawing the higher card.

The casino's edge comes from how ties are handled. If the two cards match, players may either *surrender*, forfeiting half their wager and ending the game right away, or, as in the children's game, *go to war*. If war is declared, the player must double his or her bet, and the dealer then burns three cards (removes them from play without exposing them) and deals a second card to each side. If the player's card is higher, the player wins the second bet, which is paid off at 1 to 1 odds, and the first bet is a push; if the dealer's card is higher, the player loses both bets.

A tie on the second card is handled differently at different casinos. A common resolution is to pay the player a bonus equal to the amount of the original bet while the actual bets are declared pushes, a net win of 1 unit. On a $1 bet, then, we have the following player outcomes:

Result	Win/Loss
Win without war	+$1
Lose without war	−$1
Surrender	−$.50
Win after war	+$1
Lose after war	−$2
Tie after war	+$1

The source of the casino's advantage is clear from this chart: In the event of a casino win after a tie, 2 units are won, while the player can never win more than the amount of his or her original wager even if 2 units are at risk in a war.

Example 5.3.4. Casino War also offers a Tie bet, which pays 10 to 1 if the first two cards match. Assuming a single hand dealt from a fresh shoe, the HA for this bet depends on the number of decks in play. We shall compute this house edge for a variable number of decks, denoted by n; substitution of commonly used values of n will then give the HA appropriate to any particular game.

Once the first card is drawn, we seek the number of cards remaining in the shoe that match its rank. In an n-deck game, there are $4n - 1$ such cards. Since the entire shoe holds $52n$ cards and one (the player's card) has been removed, the probability of a match is

$$p = \frac{4n - 1}{52n - 1}.$$

For a single-deck game, $p = \frac{1}{17}$; as n increases, this probability approaches

$\frac{4}{52}$, or $\frac{1}{13}$. The casino's advantage comes from paying 10 to 1 on what is no better than a 12 to 1 shot. The exact HA may be found from the equation

$$E(n) = (10) \cdot \left(\frac{4n-1}{52n-1} \right) + (-1) \cdot \left(1 - \frac{4n-1}{52n-1} \right) = -\frac{8n+10}{52n-1}.$$

The expected values are tabulated for the various values of n typically used in casinos in Table 5.2.

TABLE 5.2: House edge on the Tie bet for Casino War

# of decks	HA
1	35.29%
2	25.24%
4	20.29%
6	18.65%
8	17.83%

It is clear that you should not make this bet, but it is of interest to notice that the casino's edge actually decreases with the number of decks in the shoe, which is the opposite of the effect in a game like blackjack. Increasing the number of decks past eight (the maximum typically in use in casinos) cannot, however, eliminate the casino advantage completely. In the limit as $n \to \infty$, the expectation approaches

$$E = (10) \cdot \left(\frac{1}{13} \right) + (-1) \cdot \left(\frac{12}{13} \right) = -\frac{2}{13},$$

which corresponds to approximately a 15.38% house edge.

Example 5.3.5. It may be stretching the definition of the word to speak of a "strategy" for a game as simple as Casino War, but the question of whether to surrender or go to war after a tie is a place where player choice is involved, and thus a place where an optimal strategy may be determined.

If you surrender a $1 bet, your expectation is −$.50. If you go forward with the war, then the probability of winning is again dependent on the number of decks in use. You will either win $1, whether through winning the war or tying, or lose $2 when the war is lost. Let p be the probability of a tie; it follows that

$$P(\text{Win } \$1) = \frac{1}{2} + p$$

and

$$P(\text{Lose } \$2) = \frac{1}{2} - p.$$

We assume no knowledge about the cards remaining to be dealt, and again assume that we're starting at the top of a fresh n-deck shoe. Three cards are known at the start of the war: the two matching cards that triggered the war and the player's second card. The burn cards are not exposed when dealt, so we ignore them—the result will be another long-term average value that is suitable for quick calculations like this one. Two cases emerge:

- If your second card matches the first two in rank, then the conditional probability of a fourth card of that rank falling to the dealer is

$$p_1 = \frac{4n - 3}{52n - 3}.$$

This case has probability

$$q_1 = \frac{4n - 2}{52n - 2}.$$

- If your second card is of a different rank than the first two, then the conditional probability of a second match is

$$p_2 = \frac{4n - 1}{52n - 3},$$

and this case has probability

$$q_2 = \frac{48n}{52n - 2}.$$

The probability p of a tie and the expectation E are both functions of We find that

$$p(n) = q_1 p_1 + q_2 p_2 = \frac{208n^2 - 68n + 6}{(52n - 2)(52n - 3)},$$

where n is the number of decks in the shoe. Hence, your expectation as function of n is

$$E(n) = (1) \cdot \left(\frac{1}{2} + p(n) \right) + (-2) \cdot \left(\frac{1}{2} - p(n) \right) = \frac{-36 \cdot (26n - 3)}{(26n - 1)(52n - 3)} - \frac{7}{26}$$

For the commonly used values of n, Table 5.3 contains the expectation if you go to war. Once again, a limiting value for the case of infinitely many decks can be computed, and here that limit is $-\$\frac{7}{26} \approx -\$.269$.

Since the expected value for any number of decks is greater than $-50\cent$ follows that you should *never* surrender, and thus that the casino derives an additional advantage whenever a player surrenders.

TABLE 5.3: Expected return when going to war, Casino War

n	$E(n)$
1	-$.321
2	-$.296
4	-$.282
6	-$.278
8	-$.276

Rupert's Island Draw

Rupert Boneham, three-time contestant on the TV show *Survivor*, invented a new card game called Rupert's Island Draw in 2011. The game was subsequently offered at the Golden Nugget Casino in Las Vegas.

Rupert's Island Draw is played with a 384-card deck consisting of the aces through 6s from 16 regulation decks. Similar in some ways to blackjack and baccarat, the object of the game is to draw a hand of cards with a *lower* total than the dealer without exceeding 12. Two cards are dealt, face up, to each player and to the dealer. The action depends on the dealer's hand:

Dealer's total	Action
6, 7, 8	Dealer stands, all players get one more card.
9, 10, 11	Dealer draws a third card.
2, 3, 4, 5, 12	Each player decides who gets a third card, the player or the dealer.

The winning hand is the one with the lower total. Any hand going over 12 is revalued at 0 and is an automatic win. Ties are a push, unless player and dealer tie at 7, in which case the players lose half their bets. Winning hands pay even money [48].

The twist here is the player choice on a 2, 3, 4, or 5. (If the dealer shows a 12, a third card would put that hand over 12 and so would be a bad choice for the player.) This contrasts both with baccarat's prescribed rules for hitting or standing and with blackjack, where nothing restricts the players, but they cannot force the dealer to take another card.

Example 5.3.6. Rupert's Island Draw also offers a number of side bets on the composition of the player's and dealer's initial hands, taken together. If the hands form four of a kind, the bet pays off at 50 to 1. Working from a full shoe of 384 cards, what is the probability of this happening, and what is the expectation of this side bet?

The first card can be of any rank, but the next three must match it. We have

$$P(\text{Four of a kind}) = \frac{384}{384} \cdot \frac{63}{383} \cdot \frac{62}{382} \cdot \frac{61}{381} = \frac{39,711}{9,290,431} \approx \frac{1}{234},$$

from which we deduce that the expectation is

$$E = (50) \cdot \frac{39,711}{9,290,431} + (-1) \cdot \left(1 - \frac{39,711}{9,290,431}\right) \approx -\$.7820$$

—a 78.2% house advantage. Side bets—which, like this one, are usually easy to understand—frequently have very high house advantages, and this is one of the worst we've seen. Raising the payoff from 50 to 1 to 200 to 1 would still give the casino a healthy edge of 14.1%.

In an experimental test intended to confirm this probability, 20,000 simulated four-card hands were generated via computer. The theoretical probability derived above is approximately .004274. Our simulation produced 83 four-of-a-kind hands, for an experimental probability of .004150. This close agreement between theoretical and experimental probability suggests—but does not confirm, for no amount of data can definitively confirm theory—that our calculations are correct.

Card Slots

In 2000, Joseph R. Trucksess of New Jersey received a patent for a card game based on slot machines [92], and the game was subsequently offered at the Tropicana Casino in Atlantic City. The motivation for this game was to provide a slot machine-like game with the interaction of a table game. A card slots deck contains 52 cards that bear common slot machine symbols, distributed among the following five types (See Figure 5.4.):

- 15 Cherries (C)
- 12 Plums (P)
- 10 Lemons (L)
- 10 Oranges (O)
- 5 Triple Bars or Casino Logos (B)

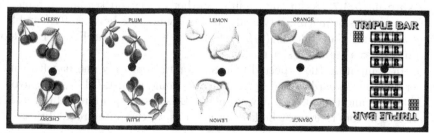

FIGURE 5.4: Card slots cards from the Tropicana Casino, Atlantic City. The cards have been canceled with a hole drilled through their centers.

The game is played like blackjack or baccarat: players make a wager, and then five cards are dealt to the table. The cards are revealed one by one, and bets are paid off depending on the cards that appear, with different combinations paying off at different multiples of the wager—just like a standard slot machine.

Several different games and payoff structures are mentioned in the patent; we shall consider one involving payoffs for certain combinations of 1, 3, 4, and 5 symbols. Using the single-letter abbreviations above, and noting that * may denote any card, the winning combinations and their payoffs, as proposed in the patent, are detailed in Table 5.4.

TABLE 5.4: Card slots payoffs [92]

1	2	3	4	5	Payoff
C	*	*	*	*	Push
C	C	C	*	*	1 to 1
P	*	*	*	*	Push
P	P	P	*	*	1 to 1
O	*	*	*	*	Push
O	O	O	*	*	2 to 1
L	L	L	*	*	5 to 1
B	B	B	*	*	25 to 1
C	C	C	C	*	50 to 1
P	P	P	P	*	100 to 1
O	O	O	O	*	200 to 1
L	L	L	L	*	400 to 1
B	B	B	B	*	500 to 1
C	C	C	C	C	250 to 1
P	P	P	P	P	500 to 1
O	O	O	O	O	1000 to 1
L	L	L	L	L	2500 to 1
B	B	B	B	B	50000 to 1

Example 5.3.7. Assuming a minimum bet of $1, and keeping in mind the fact that $1 slot machines typically return somewhere between 91% and 9 of all money they take in (4–9% HA) [66], how does the HA of card slots compare to a standard slot machine?

In computing the expected values, we note that only one payoff is made per deal of the cards—that is, if the cards come up L-L-L-L-L, the payoff is 2500 to 1, and the 400 to 1 payoff for L-L-L-L-* is not paid in addition. There are two factors that must be kept in mind as we compute probabilities:

1. The order in which the cards are dealt matters, in a way that it doesn't in draw poker. For example, the arrangement C-C-C-C-L pays 50 to 1,

while the same five cards in the order C-C-L-C-C is a push, and t order L-C-C-C-C loses. Although there are 52 cards in a card slots deck, as there are in a standard deck, there are more than $_{52}C_5$ arrangements to consider.

2. The probabilities change as cards are dealt: if the first card is a cherry (probability $\frac{15}{52}$), then the probability that the next card is a cherry is different ($\frac{14}{51}$), as are the probabilities of every other type of card once a cherry is removed.

Some of these events can be analyzed without too much mathematical difficulty. For example, in the case of the $50,000 payoff for drawing all five triple bar cards, the order doesn't matter:

$$P(\text{Win \$50,000}) = \frac{_5P_5}{_{52}P_5} = \frac{_5C_5}{_{52}C_5} = \frac{1}{2,598,960},$$

since there's only one way to pick the five cards, and all $5! = 120$ orders give the same payoff.

Some are more challenging. Consider, for example, the first winning combination listed in Table 5.4 (to the extent that a "push" bet can be a winner): C-*-*-*-*. If we work our way left to right, card by card, we get the following probability:

$$p = \underbrace{\frac{15}{52} \cdot 1}_{P(\text{C-*-*-*-*})} - \underbrace{\left(\frac{15}{52} \cdot \frac{14}{51} \cdot \frac{13}{50} \cdot 1\right)}_{\text{Remove C-C-C-*-*}},$$

where we have computed the probability of drawing a cherry as the first card and then subtracted the probability that such a hand starts with three or more cherries, which pays differently. Factoring out common terms gives

$$p = \frac{15}{52} \cdot \left(1 - \frac{14}{51} \cdot \frac{13}{50}\right) = \frac{296}{1105} \approx .2679.$$

While this is a fairly complicated calculation, it can be used as a prototype for the rest of the zero-payoff calculations. For P-*-*-*-*, we have

$$p = \underbrace{\frac{12}{52} \cdot 1}_{P(\text{P-*-*-*-*})} - \underbrace{\left(\frac{12}{52} \cdot \frac{11}{51} \cdot \frac{10}{50} \cdot 1\right)}_{\text{Remove P-P-P-*-*}} = \frac{244}{1105} \approx .2208,$$

and O-*-*-*-* leads to

$$p = \underbrace{\frac{10}{52} \cdot 1}_{P(\text{O-*-*-*-*})} - \underbrace{\left(\frac{10}{52} \cdot \frac{9}{51} \cdot \frac{8}{50} \cdot 1\right)}_{\text{Remove O-O-O-*-*}} = \frac{413}{2210} \approx .1869.$$

Adding these three probabilities together gives $P(\text{Push}) = \dfrac{1493}{2210} \approx .6756$
As a quick check on the reasonableness of this value, we note that the probability of drawing a cherry, plum, or orange as the first card is $\frac{37}{52} \approx .7115$, and that this probability includes hands that start with 3 to 5 of the same fruit and thus have a positive return. While this is not confirmation that our value is correct, it is evidence that our answer is not obviously wrong.

We can use this method to compute the probability of a losing hand as well. All losing hands start with either a lemon or a triple bar, and none starts with three or more of those symbols. The probability of a losing hand starting with a lemon is the same as that of a hand starting with one but not three oranges, $\dfrac{413}{2210}$, since there are ten lemons and ten oranges in a card slots dec For the triple bar symbol, we have

$$P(\text{Losing hand}) = \underbrace{\frac{5}{52} \cdot 1}_{P(\text{B-*-*-*})} - \underbrace{\left(\frac{5}{52} \cdot \frac{4}{51} \cdot \frac{3}{50} \cdot 1 \right)}_{\text{Remove B-B-B-*-*}} = \frac{423}{4420} \approx .0957.$$

The model for the four payouts from \$1 to \$25 is

(3 matching cards)(4th card not matching the first three)(Any 5th card).

If k is the number of cards in the deck with a given symbol X and $Y \neq X$, we have

$$P(\text{X-X-X-Y-*}) = \left(\frac{{}_kP_3}{{}_{52}P_3} \right) \cdot \left(\frac{52 - k}{49} \right) \cdot \left(\frac{48}{48} \right),$$

which gives the following probabilities:

X	k	P(X-X-X-Y-*)
C	15	37/2380
P	12	88/10,829
O	10	36/7735
L	10	36/7735
B	5	47/108,290

Collecting all of the possible payoffs and their probabilities together gives Table 5.5, the complete table of outcomes [92].

One result which can be quickly derived from this table of probabilities is that approximately 95.8% of all card slots hands result in a push or a loss. Of more interest, of course, is the expected value of a \$1 bet at card slots, which turns out to be nearly \$1.35.

A game with a 135% player advantage is clearly not sustainable on a casino floor. The Atlantic City Tropicana, we can be sure, wasn't using these payoffs.

TABLE 5.5: Card slots payoff probabilities

Payoff	Probability
0	1493/2210
1	5127/216,580
2	36/7735
5	36/7735
25	47/108,290
50	37/9520
100	33/21,658
200	3/4420
250	11/9520
400	3/4420
500	839/2,598,960
1000	3/30,940
2500	3/30,940
50,000	1/2,598,960
−1	1249/4420

Side Bets

In Section 4.2, we looked at Let It Ride. The Soaring Eagle Casino's LIR rack card describes an optional bonus wager:

> After placing three equal bets, you may add a dollar to "light up" the bonus spot. That automatically enters you into a chance for BONUS payments for a hand of "Three-Of-A-Kind" or better.

The bonus spot is a light on the table, and alerts the dealer and the other players as to who has made this bet. (The fact that the spot lights up may be a subtle form of peer pressure intended to encourage players to make this bet by making it very clear to everyone who has wagered $1 on this.) This bonus bet pays off the following amounts for the player's final five-card hand:

Player hand	Payoff	Number of hands
Royal flush	$25,000	4
Straight flush	$2500	36
Four of a kind	$400	624
Full house	$200	3744
Flush	$50	5108
Straight	$25	10,200
Three of a kind	$5	54,912

Example 5.3.8. What's the expected value of this bonus wager? Dividing each entry in the last column above by 2,598,960 gives the probability of each

winning hand. Multiplying these probabilities by the outcomes and adding the products, including the product $(-1) \cdot \left(\dfrac{2,524,332}{2,598,960} \right)$ that accounts for the losing hands, gives

$$E = -\$ \frac{550,972}{2,598,960} \approx -\$.2120.$$

Even though it's only a \$1 bet, the house edge is 21.2%.

Note from the calculation above that 2,524,332 of the possible 2,598,960 hands lose this bet—that's 97.1%.

Example 5.3.9. Another optional Let It Ride bonus bet offered at the Soaring Eagle is the "3 Card Bonus" side bet, which pays off if a player's first three cards contain at least a pair, in accordance with the following table.

Player hand	Payoff odds
Mini royal (suited AKQ)	50 to 1
Straight flush	40 to 1
Three of a kind	30 to 1
Straight	5 to 1
Flush	4 to 1
Pair	1 to 1

Note that the relative likelihood of some hands changes when we cut back from five to three cards; three of a kind is much rarer than a straight or a flush, and straights are less common than flushes. With $_{52}C_3 = 22,100$ possible hands as our denominator, the probabilities of the various three-card hands are these:

Player hand	Probability
Mini royal	4/22,100
Straight flush	44/22,100
Three of a kind	52/22,100
Straight	724/22,100
Flush	1096/22,100
Pair	3744/22,100

The probability of winning anything on the bonus bet is the sum of the six probabilities in the table: $\dfrac{5664}{22,100} \approx .2563$. The expectation on a \$1 bet is

$$E = (50) \cdot \left(\frac{4}{22,100} \right) + (40) \cdot \left(\frac{44}{22,100} \right) + (30) \cdot \left(\frac{52}{22,100} \right) + (5) \cdot \left(\frac{724}{22,100} \right)$$

$$+ (4) \cdot \left(\frac{1096}{22,100} \right) + (1) \cdot \left(\frac{3744}{22,100} \right) + (-1) \cdot \left(\frac{16,436}{22,100} \right)$$

$$= -\frac{1168}{22,100}$$

$$\approx -\$.0529.$$

The HA of only 5.29% is actually quite reasonable for bonus bets, and especially reasonable in comparison to the five-card bonus bet offered at the same table. Nonetheless, it's not low enough to recommend this wager.

Example 5.3.10. In 2011, Caesars Entertainment introduced a million-dolla Three Card Poker side bet across its American casinos: the "6 Card Bonus." With a wager of $15, this bet pays off one million dollars if the player's hand and the dealer's hand, taken together, form a "Super Royal Flush" (AKQJ109) in diamonds. The bet had hit six times by July 2012 [90].

What is the probability of winning this million-dollar prize? All that matters is that the six cards comprising the bettor's and dealer's hands are the 9 through ace of diamonds; how the cards are distributed between the two hands does not matter. We are looking at the number of combinations of 6 cards from 52, and only one of them pays off with $1 million. Accordingly, the probability of winning is

$$p = \frac{1}{{}_{52}C_6} = \frac{1}{20,358,520}.$$

Example 5.3.11. Some baccarat tables once featured the *Banker 9* side bet, a separate wager that paid off at 9 to 1 if the Banker hand was a natural 9. In [89], Edward Thorp and William Walden reported that the probability of winning this wager was .0949 in an eight-deck game. The corresponding expectation is

$$E = (9) \cdot (.0949) + (-1) \cdot (1 - .0949) = -.0510,$$

which gives the casino a 5.10% edge.

Further exploration of the banker 9 side bet may be found in Chapter 6.

Wild Cards

A common feature in casual poker games, and an occasional component of some video poker machines, is the use of *wild cards*, one or more cards in a deck that may be redefined as any card the holder desires. Common choices for wild cards are a joker added to the deck, deuces, and one-eyed jacks (the $J\heartsuit$ and $J\spadesuit$).

The incorporation of wild cards into poker has a number of effects on the probabilities of the game. One possibly unintentional effect is that, in games with a wild card, three of a kind becomes more common than two pairs. The reason for this is contained in the relative rankings of the two hands: if a player is dealt, say, $A\spadesuit$ $A\diamondsuit$ $5\diamondsuit$ $2\diamondsuit$ and a wild joker, the joker could be regarded as a third ace or as a second 5, and counting the hand as three aces is better than calling it two pairs. In short, there are no two-pair hands that include a wild card.

Example 5.3.12. Suppose we add a single joker to the deck of 52 to function as a wild card. A hand that does not contain a joker will be called a *natural* hand to distinguish it from a hand using a wild joker. The number of natural two-pair hands (see Example 2.4.15) is

$$_{13}C_2 \cdot (_4C_2)^2 \cdot 44 = 123,552.$$

This number will not increase with a wild joker, for the reason given above. Let us count next the number of three-of-a-kinds without a wild card. The rank of the triple may be chosen in 13 ways, and the three cards comprising the triple may be chosen in $_4C_3 = 4$ ways, for a total of 52 triples. Another way to think of this is that there's one three-of-a-kind corresponding to every card in the deck: the other three cards of that rank.

The two remaining cards must not match either the triple or each other; we may choose them in $(48 \cdot 44)/2$ ways. We divide by 2 because, again, the order in which the two odd cards are chosen does not matter: if we have three kings, a jack, and a 7, whether we pick the jack first and then the 7, or the other way around, the resulting hand is the same.

The total number of natural three-of-a-kinds is thus

$$\frac{52 \cdot 48 \cdot 44}{2} = 54,912.$$

To this we must add the number of three-of-a-kinds which include a joker: these hands are of the form $AABC*$, where the * denotes the joker. The joker can be dealt in only one way, so we are simply counting the number of four-card one-pair hands here. There are

$$_{13}C_1 \cdot {_4C_2} \cdot {_{12}C_2} \cdot (_4C_1)^2 = 82,368$$

three of a kinds including a joker. Adding these to the natural three-of-a-kinds gives 137,280 three-of-a-kinds in a 53-card deck with one wild joker—more than the number of two-pair hands.

In theory, then, the ranking of hands in 53-card poker should be rearranged to place two pairs above three-of-a-kind. In practice, of course, this would simply result in players opting to define a hand including a pair and a joker as two pairs rather than three-of-a-kind—and so negate the mathematical correctness of the new hand ordering. John Scarne suggested the following two possible changes to the rules of poker, for players interested in strict adherence to mathematical laws [60, p. 586]:

1. In a hand consisting of a pair plus a joker, the joker cannot be called wild and must be valued as a third unmatched card.

2. Alternately, in a hand with pair of 8s through aces plus a joker, the joker may be used to raise the hand to three-of-a-kind. A pair of 2s through 7s plus a joker must be called two pairs.

Under either one of these alternate hand-ranking schemes, three-of-a-kind is less common than two pairs. In suggesting these rule changes, Scarne was well aware that his hybrid schemes might be mathematically correct, but would be very unlikely to catch on for casual play. The first scheme simply eliminates all two-pair and three-of-a-kind hands containing a wild card. Six ranks of pairs cannot be raised to three-of-a-kind by a joker in the second scheme, while seven ranks can. This discrepancy does not overcome the excess of natural two-pair hands over natural three-of-a-kinds, and so the order of the two hands is preserved.

A later examination of the mathematics, considering one, two, or four wild cards, showed that if wild cards are allowed in poker, it is not possible to rank the hands so that more valuable hands occur less frequently without restrictions on how wild cards may be valued [11]. No matter how the various hands are ranked, there will always be a way to value hands containing wild cards so that a more common hand is ranked higher than one that is less common.

A variation on this idea may be found on some video poker machines, where a royal flush with wild cards pays off somewhat less than a natural royal flush, but video poker has the advantage of being automatically scored and paid off, and unlike live poker need not be concerned with human interpretation of its results. One paytable for a video poker machine dealing from a 53-card deck including one wild joker is Table 5.6 [35, p. 156].

The big payoff—indeed, the only inflated payoff at the five-coin level—remains the natural royal flush. A royal flush with a joker pays well, but not nearly as well as the natural royal, and it collects no bonus when betting max coins. An additional modification that appears in this table is the fact that two pairs is now the lowest paying hand, although in the play of the game, a pair plus a joker is interpreted as three-of-a-kind. Some joker-wild games pay off on a pair of kings or better; this usually comes along with slightly lower payoffs further up the table.

Example 5.3.13. An obvious question arises: If you are dealt a joker royal flush, is it a good idea to discard the joker and go for the natural royal? The

TABLE 5.6: Video poker paytable with one wild joker

Poker Hand	Payoff: 1 coin	Payoff: 5 coins
Natural royal flush	500	4700
Five of a kind	100	500
Joker royal flush	50	250
Straight flush	50	250
Four of a kind	20	100
Full house	8	40
Flush	7	35
Straight	6	30
Three of a kind	2	10
Two pair	1	5

probability of hitting it is 1/48, but the payoff is greater by a factor of 18.8 when playing max coins, and there are smaller payoffs possible to fill the gap.

This problem is similar to the one explored in Example 4.2.29, with a similar table of probabilities. Assume that you have been dealt $KQJ10$ along with a joker. On a \$1 machine with max coins (\$5) bet, your possible outcomes if you discard the joker are these:

Result	Net Payoff	Probability
Natural royal flush	4695	1/48
Straight flush	245	1/48
Flush	30	7/48
Straight	25	6/48
Nothing	−5	33/48

The expectation if you hold the joker is \$245. If you discard the joker, your expectation is

$$E = (4695) \cdot \frac{1}{48} + (245) \cdot \frac{1}{48} + (30) \cdot \frac{7}{48} + (25) \cdot \frac{6}{48} + (-5) \cdot \frac{33}{48}$$
$$= \frac{5135}{48}$$
$$\approx \$106.98.$$

It's not even close. As one might expect, breaking up a joker royal flush is not a good strategy, and the corresponding strategy table for this game [35, p. 211–2] makes that clear: A joker should never be discarded.

A second option seen in video poker is "Deuces Wild," in which all four 2s function as wild cards. In the long run, of course, this leads to higher-ranked hands, and the paytable must accommodate that. On a Deuces Wild machine, payoffs for a pair of jacks or better and for two pairs are typically eliminated, and the lowest hand that pays off is three-of-a-kind.

Example 5.3.14. There are four natural royal flushes. In each hand, we can replace up to four cards by 2s and still have a royal flush. What is the probability of a royal flush in Deuces Wild poker?

We proceed by counting the number of 2s. If there is one deuce in the hand, there are five cards that it can replace, and four choices for the replacing card. Multiplying by the four natural royal flushes gives a total of $4 \cdot 5 \cdot 4 = 80$ one-deuce royals.

If our royal flush contains two deuces, we can pick the deuces in $_4C_2$ ways, and the exiting cards in $_5C_2 = 10$ ways. Across all four natural royals, this leads to $4 \cdot 6 \cdot 10 = 240$ royal flushes with two wild cards.

A three-deuce royal has four choices for the deuces and $_5C_3 = 10$ ways to pick the replaced cards. There are therefore $4 \cdot 4 \cdot 10 = 160$ such hands.

Four-deuce royals are easy to count: We simply combine the four deuces with any of the 20 cards that can be part of a royal flush, for a total of 20.

Adding everything up gives a total of 504 royal flushes, 126 times the number without wild cards, and so the probability of a royal flush is also 126 times greater, or

$$\frac{504}{2,598,960} = \frac{3}{15,470} \approx 1.939 \times 10^{-4}.$$

This plethora of new royal flushes is reflected in the paytables, where a natural royal flush pays anywhere from 10 to 12.5 times as much as a royal flush with one or more deuces, depending on the exact game being played [35, p. 149–154]. Of course, each paytable carries its own perfect strategy, and players using the wrong strategy will increase the often-tiny house advantage.

For a look at an experiment with wild jokers in blackjack, see Example 6.3.3.

5.4   CASINO PROMOTIONS

We have seen the Baldini's promotion (Examples 2.4.6 and 4.2.4) of a free spin on a video poker machine as an incentive for local patrons to cash their paychecks at the casino. Another incentive for players comes in the form of *matchplay* coupons, which allow players to make a bet at a table game and have the amount they wager matched (in principle; no additional chips are played) by the casino. The net effect of a matchplay coupon is to double the payoff, should the bet win. Typically, these coupons are restricted to even-money wagers and have some maximum bet limit; for example, a matchplay coupon offered by the Grand Lodge Casino of Incline Village, Nevada effectively offers a 2 to 1 payment on any even money bet of $10 or less by doubling the amount of a player's wager. The coupon must be played with a live bet of the amount indicated and is surrendered after the bet is either won or lost; in the case of a tied bet, as in blackjack, the player may re-bet the coupon. It is not unusual for this increased payoff to tip the game to the player's advantage, hence the bet limit and restrictions on the coupon's use.

Example 5.4.1. For a $10 matchplay bet on red at roulette, what is the player's advantage?

The probability of winning has not changed; what is different is the payoff to a winning player, which has doubled. In American roulette, we have

$$E = (20) \cdot \left(\frac{18}{38}\right) + (-10) \cdot \left(\frac{20}{38}\right) = \frac{160}{38} \approx \$4.21,$$

which, when divided by the $10 initial bet, gives the player a 42.1% edge over the casino.

Example 5.4.2. If this coupon could be used on a 35 to 1 single number bet in roulette, what would the player's edge on a $10 wager be?

In this case, a payoff of $350 would be doubled to $700, and so

$$E = (700) \cdot \left(\frac{1}{38}\right) + (-10) \cdot \left(\frac{37}{38}\right) = \frac{330}{38} \approx \$8.68,$$

for a player advantage of 86.8%.

Related to matchplay coupons are *nonnegotiable* or *no cash value* (NCV) chips. These special chips have a face value, but may not be redeemed at the casino for cash. Instead, they must be played at the tables at least once. They need not be accompanied by matched cash bets, and if a bet made with these chips wins, the player is paid with ordinary negotiable casino chips. NCV chips come in two types. The first type, and the most common, may be wagered only once, and like a coupon, is taken after a resolved bet—win or lose. The second type, while occasionally offered to any casino patron, is usually available only to a casino's best (read: richest) customers, and can be retained after a win to bet again, until they lose.

Both forms of nonnegotiable chips are, like matchplay coupons, typically restricted to even-money wagers. One-use chips have an expected value of approximately half their face value, because they are only in play for a single wager. If an NCV chip has a face value of $A and is wagered on an even-money bet whose probability of winning is $p < .5$, it follows that

$$E = (A) \cdot p + (0) \cdot (1 - p) = A \cdot p \approx \frac{A}{2},$$

where the expectation is roughly as close to $A/2$ as p is to $\frac{1}{2}$. For even-money bets, p is usually very close to, but always less than, $\frac{1}{2}$.

Bob Stupak took this consideration and turned it around on his customers at Vegas World. Stupak was known in Las Vegas for his "VIP Vacation" packages, which he liberally offered and aggressively marketed in order to attract visitors to Vegas World, which was located on Las Vegas Boulevard about half a mile north of Sahara Avenue, the traditional northern boundary of the Las Vegas Strip, in a neighborhood of questionable reputation known as "Naked City." The packages offered a low-cost Las Vegas vacation that included two nights at the resort, complimentary drinks, and free gambling money in the form of cash and special gambling chips, as much as $1000 in chips per visit, all for less than the face value of the chips alone. These chips, not quite NCV chips, were inscribed with a table value and a cash redemption value of $\frac{1}{4}$ that value, so a $25 chip had a cash value of $6.25 and a $5 chip had a cash value of $1.25. Vegas World guests who were unfamiliar with gambling mathematics and were happy to cash in their chips for their redemption value were handing the casino nearly 25% of the chips' face value, for, as was shown above, the chips had an expected return of close to half their face value.

An alternate Vegas World promotion issued patrons two $100 bills when they checked in, with instructions to take them to the casino cashier to exchange them for $600 in nonnegotiable chips [77]. Customers, of course, were free to disregard the instructions and keep the money; anyone who did so was essentially forfeiting nearly $100 in value. Many did.

Example 5.4.3. What would be a good way to use those Vegas World chips? As we will see in Section 7.6, maximizing your return in a casino is best done by making a few large bets rather than many small ones. Several people independently made the following discovery: if you have $1000 in non-negotiable chips, take them to a roulette table and bet $500 on red, bet the other $ on black, and put $20 in regular casino chips on each of the zeroes. You have every possibility covered, and so cannot lose. The outcomes of this wagering combination are these:

- **A red or black number comes up.** You lose your bets on the zeroes (−$40), but win $500 in casino chips on your color bet. Your NCV chips are taken, but you can walk away with a $460 profit.

- **0 or 00 is spun.** You lose all of your NCV chips, but win $700 on your winning bet, from which we subtract the $20 you wagered on the losing number, for a net profit of $680.

Your expected return with this combination bet at an American roulette table is

$$E = \underbrace{(460) \cdot \left(\frac{36}{38}\right)}_{\text{Red or black number appears}} + \underbrace{(680) \cdot \left(\frac{2}{38}\right)}_{\text{0 or 00 appears}} \approx \$471.58.$$

This is a guaranteed return of about 99.56% of the $473.68 (1000 × theoretical expected value of $1000 in NCV chips when wagered on an even-money roulette bet.

Example 5.4.4. It's possible to do better. Take your $1000 to a craps table and put $500 on pass, $500 on don't pass, and $20 cash on 12, a one-roll bet paying 30 to 1 if a 12 is rolled. As with the roulette example, every possible roll of the dice leads to a win for you.

- If a 12 is rolled on the come-out roll, your pass line bet chips are taken, but your don't pass chips are still yours to bet again, and your bet on 12 wins $600.

- In any other case, exactly one of your pass and don't pass bets wins, whether by a 2, 3, 7, or 11 on the come-out roll or by the shooter setting and making a point. You win $500 but lose the $20 wager on the 12, for a profit of $480 to take away with you.

Your expectation on the first bet with this wagering strategy is

$$E = (480) \cdot \left(\frac{35}{36}\right) + (600) \cdot \left(\frac{1}{36}\right) \approx \$483.33.$$

The theoretical expectation of $1000 bet this way is $485.50; your return here is again 99.56% of the long-term average return. If the come-out roll was a 12, you can continue by mimicking this wager with the $500 in NCV chips you originally bet on don't pass, which you retain since that bet was a tie.

Casino personnel, while not happy to see the NCV chips being used in this way, evidently made no move to stop players from making these bet combinations. Of course, most people were not aware of this potential and made many bets against the intractable house edge.

Example 5.4.5. A nonnegotiable chip that can be re-bet until it loses has an expected value of approximately its face value. This can be seen in the following calculation: Assume once again that the chip has a face value of $ and is being wagered on a bet where the probability of winning is $p < .$ repeatedly until it loses. Since the chip cannot be cashed in, this is standard procedure with this type of NCV chip. We then have

$$E = (0) \cdot (1-p) + (A) \cdot p \cdot (1-p) + (2A) \cdot p^2 \cdot (1-p) + \cdots$$
$$= \sum_{k=0}^{\infty} kA \cdot p^k \cdot (1-p)$$
$$= A \cdot (1-p) \cdot \sum_{k=0}^{\infty} k \cdot p^k$$
$$= A \cdot (1-p) \cdot \left(\frac{p}{(p-1)^2}\right), \quad \text{as seen in Example 5.2.2.}$$
$$= \frac{A \cdot p}{1-p}.$$

It should be noted that this calculation includes some very unlikely events with very high payoffs, such as winning 1000 straight bets. If $p = .5$, then the expectation above is A; if $p < .5$, then the expectation is less than A. For most even-money casino bets, p is less than, but close to, .5, and so an upper bound for the value of a replayable nonnegotiable chip is its face value.

Table 5.7 collects some common even-money wagers and shows the expectation of a renewable NCV chip when used on these propositions.

TABLE 5.7: Expectation of an $A renewable NCV chip

Wager	P(Win)	Expectation
American roulette, even money bet	.4737	$.9000 \cdot A$
Blackjack, basic rules	.4750	$.9048 \cdot A$
Craps, don't pass line	.4790	$.9194 \cdot A$
European roulette, even money bet	.4865	$.9474 \cdot A$
Craps, pass line	.4920	$.9685 \cdot A$
Baccarat, player	.4932	$.9732 \cdot A$

Other casino promotions come in the form of games that offer a particularly advantageous wager, but without a matchplay component.

Example 5.4.6. A standard $3 bet on a six-spot keno ticket at Michigan's Kewadin Casinos has the following payoff table:

Outcome	Payoff (for 1)	Net win
Match 3	$3	$0
Match 4	$12	$9
Match 5	$270	$287
Match 6	$4500	$4497

In 1990, the Continental Casino in Las Vegas (now the Silver Sevens) offered a six-spot keno game with the following pay structure for a minimum $3 bet [99]:

Outcome	Payoff (for 1)	Net win
Match 4	$15	$12
Match 5	$800	$797
Match 6	$5500	$5497

This game sacrifices the break-even payment on a "match 3" outcome and offers higher payoffs for the less-likely wins. The expected return on a $3 bet at this game is

$$E = \left[15 \cdot \frac{{}_6C_4 \cdot {}_{74}C_{16}}{{}_{80}C_{20}} + 800 \cdot \frac{{}_6C_5 \cdot {}_{74}C_{15}}{{}_{80}C_{20}} + 5500 \cdot \frac{{}_6C_6 \cdot {}_{74}C_{14}}{{}_{80}C_{20}} \right] - 3$$

$$= \frac{194,217}{316,316}$$

$$\approx \$.614,$$

a positive expectation which gives a player edge of 20.47%.

A challenge in fully exploiting this opportunity is that the probability of even a "match 4" win is only about 2.85%, and so there is some risk that your bankroll might not be large enough to continue betting until you win.

Example 5.4.7. In December 1989, the Sahara Casino in Las Vegas ran a baccarat promotion that eliminated the commission entirely [99]. The expectation of a $1 Banker bet was then

$$E = (1) \cdot (.4584) + (0) \cdot (.0955) + (-1) \cdot (.4461) = \$.0123,$$

giving gamblers a 1.23% edge on Banker bets.

Although the promotion ran only from Monday through Thursday, and then only from 7:00 P.M. to 2:00 A.M., the attraction was so great that seats were seldom available at those tables.

5.5   EXERCISES

Answers begin on page 255.

5.1. The Hollywood Casino in Joliet, Illinois claims to be the only casino offering the "All Day 2" bet, which is a bet that a 2 will be rolled before a 7. There is also an "All Day 12" bet, which is mathematically identical to All Day 2. This bet pays off at 5 to 1. Find the house advantage.

5.2. Here's a combination of chuck-a-luck and craps that has been seen in the craps pit at a number of casinos. The player may bet on any single number from 1 to 6 and is paid off according to the number of dice that show that number. If two dice show your number, the payoff is 4 to 1; if one die shows your number, the bet is paid off at 2 to 1. Find the expected value of a $1 bet on a single number. How does this compare with the HA of more traditional craps bets?

5.3. We saw in Example 5.2.9 that switching from six-sided to eight-sided dice in chuck-a-luck increases the already high house advantage to an even higher and thus less attractive number. It stands to reason (and can be confirmed mathematically) that switching from six-sided to four-sided dice would give the players too much of an edge, but what if we used nonstandard d4's? Consider a chuck-a-luck cage containing three four-sided dice numbered as follows:

Die	Numbers
Red	1, 2, 3, 4
Green	5, 6, 1, 2
Blue	3, 4, 5, 6

Note that for any number you choose, one of the three dice cannot show it, and so the maximum payoff is \$2. Find the house advantage for a \$1 bet on a single number.

5.4. The Stratosphere Casino in Las Vegas, successor to Vegas World, continues to offer crapless craps and does so with 10X odds available.

a. Remembering that odds bets are paid off at true odds with no house advantage, find the payoff odds for an odds bet when the point is 2 or 12.

b. Find the payoff odds for an odds bet on 3 or 11.

c. Find the house advantage of a \$5 pass line bet backed up with maximum 10X odds.

5.5. The Four Winds Casino in New Buffalo, Michigan offered a dice ga called *Four the Money*, which uses some of the ideas of craps. In this game, players may bet on any two-die total from 2 through 12, other than 7. A bet wins if the player's number is rolled twice before a 7 is rolled. Bets are paid off in accordance with the following table:

Roll	2	3	4	5	6
Odds	45 to 1	14 to 1	7 to 1	5 to 1	7 to 2
Roll	8	9	10	11	12
Odds	7 to 2	5 to 1	7 to 1	14 to 1	45 to 1

Which numbers have the best expectation for the player? Notice that the symmetry of the payoff table makes it possible to cut the work required here in half.

5.6. The "Four for the Money" side bet in Four the Money offers even money if no 7s are rolled in four tosses of the dice. Find the expectation of this bet.

5.7. Suppose that a casino removes the zeros from its roulette wheel, leaving only 36 pockets and pays 34 to 1 on a single-number bet. How does the expected value of a \$1 bet on this new wheel compare to the previous expectations? Consider both American and European wheels.

5.8. Confirm the assertion in Example 5.1 that the house advantage for a bet that the number 23 will come up on both wheels of a European Double Action roulette game is 12.3%.

5.9. *Multicolore* is a roulette-like game that debuted at Monte Carlo in the 1950s [12, p. 115–6]. A wheel is divided into 25 sectors: six each colored green, red, white, and yellow, and one colored blue. A billiard ball is tossed into the spinning wheel and eventually comes to rest within one of the sectors. Players bet on a color: the blue sector is labeled 24 and the six sectors of each of the other four colors are labeled 4-3-3-3-2-2. A bet on blue pays off at 24 to 1, a bet on any other color pays off at x to 1 in accordance with the number x the sector.

a. Find the house advantage for a bet on blue.

b. Find the house advantage for a bet on any of the other colors.

c. What is the best betting strategy for Multicolore?

5.10. *Boule* or *La Boule* is another roulette variation that, as the name suggests, is of French origin [56]. The wheel is much simpler, using only the numbers 1 to 9. A rubber ball is thrown into the spinning wheel and eventually settles into a hole at the center. Each number has four holes allocated to it. Numbers 1, 3, 6, and 8 are black; 2, 4, 7, and 9 are red. The 5 is yellow and functions somewhat like the green zero in standard roulette. The following bets are available at La Boule:

Bet	Payoff	Description
Red/Black	1 to 1	Bet on red or black
Low (Manqué)	1 to 1	Bet on 1, 2, 3, 4
High (Passé)	1 to 1	Bet on 6, 7, 8, 9
Odd/Even	1 to 1	Bet on odd or even
Single number	7 to 1	Bet on any one number

The number 5 is neither odd nor even, high nor low. Notice that payoff structure of La Boule is considerably simpler than standard roulette or Royal Roulette—only two different payoffs are available. Compute the house advantage on each of the two payoffs. How do these HAs compare to those of standard roulette and Royal Roulette?

5.11. At the same time that Diamond Roulette was approved in New Jer state gaming officials also approved a *seven-number* bet, which must be made on the block of numbers 10, 11, 12, 13, 14, 15, and 33. If one of those seven numbers is spun, the bet pays off at 4 to 1 [50]. Find the expected value of a $1 seven-number bet on both single-zero and double-zero wheels.

5.12. The white neighborhood bet in Riverboat Roulette is a one-spin wager that pays off at 7 to 2: if a number of any other color is spun, the bet loses. Find the expectation on a $2 bet on white.

5.13. Another sic bo bet is the *Big* bet, which pays off at 1 to 1 if the sum of the three dice is between 11 and 17 inclusive, with the provision that the bet loses if triples are rolled. Find the HA of the Big bet.

5.14. The *Any Triple* sic bo bet pays off at 30 to 1 if all three dice show the same number. What is the house advantage of this bet?

5.15. Harrah's Casino in Reno, Nevada offers two different Pick 3 Keno games. The first pays off at 1 for 1 if two of your three numbers are drawn and 41 for 1 if all three are drawn. The second, called "Catch-All," pays nothing if only two numbers are drawn, but pays off at the higher rate of 128 for 2.50 (the required wager is $2.50) if all three are drawn. Which of these games is a better bet for the player?

5.16. Another Vegas World promotion was a $100 "Field Roll" chip. Find the expectation of this chip when used on a field bet (see page 85). Assume that the chip is taken after the field bet is resolved.

Blackjack: The Mathematical Exception

Thus far, we have not said much about *blackjack* or *21*, arguably the most popular casino table game and certainly one of the most popular subjects of casino mathematics study. There are good reasons for this—some mathematical, some logistical:

- As mentioned in Section 3.2, successive hands of blackjack are *not* dependent, and so the Multiplication Rule in its simplest form is not applicable. In part because of this nonindependence, the probability of winning a hand and the expected value of a bet are somewhat mor difficult to calculate.

- The rules of blackjack fluctuate from casino to casino—or even from table to table within a casino, making universal calculations impossible. While the best that we can do is build a collection of probabilities as are appropriate for different game conditions, this is no barrier to interesting mathematics. It simply means that we must be careful about tailoring our calculations to the game at hand and specifying the applicability of our conclusions.

- Blackjack is a game with a skill component, and so player action and player error must be factored into any calculations.

  A novice blackjack player would do well to learn *basic strategy* (Section 6.3), which is a set of rules for decision making for a player to follow. Based on millions of simulated hands, basic strategy gives a player the best advice on how to play a hand, given the composition of the hand and the dealer's upcard. Calculations about the HA typically assume that the player is using basic strategy correctly; deviations from these instructions will work to increase the casino's long-term edge.

6.1   RULES OF BLACKJACK

Contrary to many people's understanding, the object of blackjack is not to get a hand of cards totaling as close to 21 as possible without going over. Rather, the object is to get a hand that is *closer to 21 than the dealer's hand* without going over.

If the first object listed were the point of the game, stopping on a hand of 12, as is often called for in blackjack basic strategy, would be a bad idea. However, there are times when standing on a relatively weak hand in hopes that the dealer's hand will "bust," or go over 21, is the best strategy for a player.

Basic Rules

Blackjack uses anywhere from one to eight decks of cards shuffled together. Two cards are dealt to each player, and two—one face up (as in Caribbean Stud Poker, this is called the *upcard*) and the other face down (the *hole card* to the dealer. Players' cards are customarily dealt face down from the dealer's hand in single- or double-deck games, and face up from a shoe in games using four or more decks. Each card counts its face value, with the exceptions that face cards count ten, and an ace may be counted as either 1 or 11, at the player's discretion. (In writing about blackjack, it is common to use the shorthand "ten" to refer to any card—10, jack, queen, or king—counting as 10.)

A hand containing an ace counted as 11 is called a *soft* hand, because it cannot be busted by a one-card draw—if a hand counting an ace as 11 goes over 21 upon drawing a card, the ace may simply be revalued at 1. A hand with a total of 12 to 16 without any aces, or with all aces counted as 1, is called a *hard* or *stiff* hand, because drawing a single card risks busting the hand. A player dealt a two-card total of 21 consisting of an ace and a ten-count card—called a *natural* or *blackjack*—wins immediately until the dealer also has a natural. Naturals pay either 3 to 2 or 6 to 5, depending on the rules of the casino. (The alternate name "blackjack" for a two-card total of 21 derives from the earliest days of the game, when a 21 consisting of the ace and jack of spades qualified for a bonus payoff.)

If the dealer does not have a natural, then each player in turn has the opportunity to "hit" their hand and take additional cards in an effort to bring their total closer to 21 without going over. If a player's hand exceeds 21, this is called *busting* or *breaking*, and the bet is lost and collected at once.

Once all player hands are settled, the dealer exposes the hole card. If the dealer's hand is 16 or less, he must take additional cards. The dealer must stand on a hand of 17 or higher, although many casinos also require dealers to hit a soft 17 hand. When the dealer's hand is complete—either by busting or reaching a total of 17 or higher—the hand is compared to those of all players who have not yet busted. Player hands that are closer to 21 than the dealer's are paid off at 1 to 1; if the player and the dealer have the same total, the hand is called a *push*, and no money changes hands. If the dealer's completed hand is closer to 21 than the player's, the player loses and his or her wager is collected.

Example 6.1.1. Suppose that two players are facing the dealer in a blackjack game dealt from a single deck. Player 1 is dealt $Q\spadesuit$ 8$\clubsuit$ and player 2's hand is 6$\heartsuit$ 4$\heartsuit$. The dealer's upcard is the 3$\heartsuit$.

Player 1 chooses to stand on her total of 18, and player 2 hits his 10, drawing the 2$\heartsuit$ and bringing his total to 12. He draws a fourth card, the 10 and busts with 22, losing his bet. The dealer turns over the $K\heartsuit$ for a 13, and must draw. His third card is the 2$\diamondsuit$, bringing his total to 15. Since this is still less than 17, he draws again and receives the 2$\clubsuit$. His total is 17, and he stops. Player 1 wins with her 18.

The advantage for the casino lies in the fact that the players must play out their hands first, and if they bust, they lose even if the dealer subsequently busts with a higher total. The blackjack rule that "ties are a push and the player neither wins nor loses" applies *only* to ties at 21 or less—if a player and dealer tie with hands of 23, the player's chips are already in the dealer's rack before the dealer busts, and they aren't returned.

Additional Rules

Depending on the casino, players may be offered several options during play to make additional bets that offer the chance of winning more money (or, in the case of surrender, losing less money). These are not options available to the dealer.

- If the player's first two cards are the same—as in a pair of 8s or aces—they may be *split* to form the first card of two separate hands. The player must match his bet on the new hand, and the two hands are played out separately. Some casinos allow players to split two 10-count cards, such as a jack and queen, and a standard casino rule allows the player to draw only one additional card to each hand after splitting aces. If a third card matching the first two is drawn to a split hand, many casinos allow that hand to be split again, although some do not allow resplitting of aces. Most casinos have a limit on the number of times a given hand may be split: a maximum of four separate hands is common.

- The player has the option to *double down*—to double his or her initial bet after the first two cards are dealt. This represents a chance for the player to get more money in play upon receiving a good initial hand, but this opportunity comes at a cost: only one additional card may be drawn to a doubled hand. Candidates for double-down hands are hands totaling 9, 10, or 11, as well as certain soft hands. Casinos may place restrictions on which hands may be doubled; some, for example, restrict doubles to 10s and 11s. Some casinos do not allow players to double down after splitting pairs. It is also possible to "double down for less": to increase the bet by less than the full amount originally wagered—this is only recommended if you absolutely cannot afford to double your bet, as doubling for less means taking less than full advantage of a situation where you have the edge over the house.

- If the dealer's face-up card is an ace, players have the opportunity to make an *insurance* bet. This is a separate bet of up to half their initial wager and pays 2 to 1 if the dealer has a natural. Of course, if the dealer does have 21, the main hand loses (unless the player also has a natural), and this is the reason for the name—the player is "insuring" the main hand against a dealer 21.

- Some casinos offer a *surrender* option, in which a player may elect not to play out his hand and forfeits only 50% of his initial wager. This might be something worth considering when the chances of beating the dealer are small—for example, when the player's hand is 16 and the dealer's upcard is a 10. Surrender comes in two versions: *early surrender*, where the option is available before the dealer checks his or her hole card for a possible natural, and the far more common and less player-friendly *surrender*, which is only offered after the check for a natural is complet

In light of these options, the set of outcomes on a $1 blackjack bet can contain more than the two outcomes $\{1, -1\}$ that follow from a $1 bet on red at roulette, and this range of options leads to more complicated—and more interesting—mathematics. If the player doubles down, it's possible to win or lose $2, or to break even. A natural results in a win of $1.50. Surrendering introduces the possibility of a loss of $.50. If an initial pair is split, then it's possible to win $2, win $1, break even, lose $1, or lose $2. If pairs can be resplit, the list of possible outcomes grows.

Players may play more than one hand at most blackjack tables, provided that space is available. Most casinos, however, require that gamblers playing multiple hands bet more than the table minimum on the extra hands, frequently double the minimum on a second hand and triple the minimum on the third hand. While playing multiple hands gives no advantage in winning any particular hand, this can be a way to bet more money than the table maximum and thus win more if the dealer busts. If the hands are dealt face down, as is usually the case in one-deck and two-deck games, players are not allowed to look at the cards in one hand until any previous hands have been played out to completion.

Depending on the set of rules, the number of decks used, and the payoffs on naturals, the house advantage on a hand of blackjack, in practice, can range from essentially 0% to upwards of 3% for a player using basic strategy. (Deviations from basic strategy increase the HA.) At the outset, let us assume that we are playing single-deck blackjack under what are called "standard Las

Vegas Strip rules" [91]. This name is used out of convenience; these rules are neither exclusive to the Strip nor universal thereon:

- Dealer stands on all 17s.

- Players may double on any two cards, and may double after splitting a pair.

- Aces may be resplit.

- Naturals pay 3 to 2.

With these rules in force, single-deck blackjack is nearly an even game for a player using basic strategy. Table 6.1 (from [23] and [28]) lists the effect on the house edge of various rules changes.

TABLE 6.1: Effects of rule changes on blackjack house advantage

Rule change	Effect on HA
Two decks	+.32%
Four decks	+.48%
Six decks	+.54%
Eight decks	+.57%
Dealer hits soft 17	+.20%
No double downs on soft hands	+.13%
Double down only on 10 or 11	+.26%
No double downs on 9	+.13%
No double downs on 10	+.52%
Double downs only allowed on 11	+.78%
No double downs allowed	+1.60%
No nonace pair resplitting	+.03%
No splitting of aces	+.18%
Pair splitting forbidden	+.40%
Dealer wins ties	+9.00%
Blackjacks pay 6 to 5	+1.39%
Blackjacks pay even money	+2.32%
Double down on two or more cards	−.24%
Double down after pair splitting	−.14%
Late surrender	−.06%
Early surrender	−.62%
Resplitting Aces allowed	−.06%
Draw more than one card to split Aces	−.14%
Blackjack pays 2 to 1	−2.32%
Six cards under 21 automatically wins	−.15%

Example 6.1.2. These percentages may be combined to determine the house edge for a blackjack game with a specified set of rules. For example, consider the electronic blackjack game produced by International Game Technology and marketed as part of the Game King suite of electronic table games with the following rules:

- Six decks are used (HA +.54%)

- Blackjacks pay even money (HA +2.32%)

- No pair splitting allowed (HA + .4%)

- Doubling down not permitted (HA + 1.6%)

- Six cards under 21 automatically wins (HA −.15%)

Adding up the cumulative effects of these five rule changes results in a game with a house edge of 4.71%—which is close to the HA for American roulette, even though gamblers might think they're getting a fairer game "because it's blackjack."

Side Bets and Variations

Many new ideas and game variations have been proposed for blackjack bets. In part this is because blackjack is in part a game where skill at making choices plays a role, and so if there's a new bet offering a chance for players to make incorrect choices that favor the casino, that bet may seem attractive to some casino officials. In part this may be because blackjack using basic strategy is nearly an even game, and these side bets offer the casino an opportunity to increase its profits, as these bets frequently have a high house edge.

- The *Royal Match* bet allows players to make a side bet on their first two cards. The paytables for Royal Match vary somewhat; one version pays 2½to 1 if the first two cards are suited (the "Easy Match") and 25 to 1 if they are the king and queen of the same suit (the "Roya Match"). An additional option is to offer a progressive jackpot for the "Crown Treasure," which happens when both the player and dealer have a Royal Match.

- *Instant 18* is a separate side bet that must be made alongside a regular blackjack wager. When you make an Instant 18 bet, you are effectively playing a second hand that is assigned a value of 18. No cards are dealt to this phantom hand.

- *Wild jokers* were introduced into blackjack at the Sands and Dunes casinos in Las Vegas [98]. Three jokers were added to a six-deck shoe. A joker dealt to a player had to be assigned a card value before the hand continued; jokers dealt to dealers were burned and a new card dealt to the hand.

- Bob Stupak, whom we first met in Example 5.2.1 in connection wi
  Crapless Craps, also invented a variation of blackjack called *Double Exposure 21*, in which both of the dealer's cards are revealed to the playe
  This, of course, gives the player a large advantage, but that is more than
  compensated for by new rules [83, p. 93]:

  - The house wins all ties, except ties between naturals, which
    player wins.

    This may lead to some unusual decisions that would never be ma
    in ordinary blackjack, as when a player holding 20 against a dealer
    20 must take a hit and hope for an ace, for 20 vs. dealer 20 is a
    losing hand.

  - Naturals pay only even money, except naturals consisting of
    and J♠, which pay 2 to 1. A three-card 21 consisting of a suited 6,
    7, and 8 also pays 2 to 1.

  - Doubling down is only permitted on 9, 10, and 11.

  - Pairs may be split only once.

Double Exposure II, introduced somewhat later, exposed the dealer's
hole card only when it was a ten-count card, and forbade doubling and
splitting pairs. The payoff on a suited 6-7-8 in Double Exposure II was
triple instead of double.

- *Super Fun 21* claims to be "the most exciting way to play blackjack"
  according to the Orleans Casino's rack card for the game. In Super Fun
  21, players have the following additional options:

  - Double down on two *or more* cards, including after hitting or splitting pairs. If you have a three-card total of 11, you may doubl
    down—this is not allowed under standard blackjack rules.

  - Pairs, including aces, may be resplit up to three times, resulting in
    as many as four hands in play at a time.

  - Hands may be surrendered at any point, including after drawi
    one or more cards, if the dealer does not have a natural.

  - A hand totaling 20 or less with six or more cards pays even mone
    regardless of the dealer's hand, although if the hand has been doubled, the doubled amount is not an instant winner. (It does remain
    in play and is paid if the hand beats the dealer.)

  - A 21 with five or more cards pays 2 to 1 instantly, with the same
    restriction as for six-card 20s.

  - Player naturals are instant winners even if the dealer also has a
    natural. Naturals with both cards diamonds pay 2 to 1.

The following rule acts against the player and more than compensates
the casino for the more player-friendly rules above:

- Blackjacks other than in diamonds pay even money instead of 3 2 or 6 to 5.

- *Spanish 21*, introduced in 1995, is a blackjack variant played with six t eight decks of 48 cards from which all four 10s have been removed—this is sometimes called a "Spanish deck"[80]. Removing all of the 10s raises the casino's advantage by 2%, $\frac{1}{2}$% per removed card, but a number of rule changes are added to make up for this effect [103]. There is some variability from casino to casino, but the alternate rules usually include these:

  - Players' naturals always win, even if the dealer also has a blackjack. Blackjacks pay 3 to 2.

  - All player totals of 21 beat dealer 21s. Other ties are pushes, as is the case in regular blackjack.

  - Players may double down on any number of cards, not just the fir two.

  - Players may double down after splitting.

  - Players may resplit pairs, including aces, for up to four hands.

  - A number of bonus payoffs for certain player hands are offered—f example, a five-card 21 pays 3 to 2, a six-card 21 pays 2 to 1, and a 21 consisting of seven or more cards pays 3 to 1.

- In *Multiple Action 21*, players make two or three separate bets on their hand, and the dealer plays out his or her hand three times, each time starting with the same upcard. If the player's hand busts, all the bets are lost at once. The game was developed at the Four Queens Cas in Las Vegas, which holds a patent on the game and has licensed it to other casinos [98].

- *Blackjack Switch* calls for each player to play two hands, with separate wagers of the same size on each. After the cards are dealt and the dealer has checked for a natural, players then have the option to switch the second cards in each hand from one hand to the other in order to create more favorable totals.

  Example 6.1.3. If a Blackjack Switch player is dealt $K\clubsuit$ $5\diamondsuit$ on one hand and $6\spadesuit$ $Q\spadesuit$ on the other, he or she may rearrange the cards to $K\clubsuit$ $Q\spadesuit$ and $6\spadesuit$ $5\diamondsuit$, thus changing two mediocre hands of 15 and 16 to a strong 20 and an 11. The player may then double down on the 11.

  The dealer plays only one hand and thus does not have the advantage given by switching cards. The player advantage is tempered by the following two rule changes:

- All blackjacks pay even money rather than 3-2 or 6-5. Typically, a switched Ace-10 counts as a 21 rather than a blackjack, and so not paid off immediately because it may be tied later by a deale 21.

- If the dealer's hand totals a hard 22, then all remaining bets, including any 21's arising from switching cards, push. This rule turns a losing dealer hand into an automatic tie. A player 22 remain losing hand, even if it's later tied by a dealer hard 22. (A dealer's soft 22 would be revalued at 12, and additional cards drawn.)

Other rules of blackjack—doubling down, splitting pairs, and insurance—remain as in standard blackjack.

6.2 MATHEMATICS OF BLACKJACK

Some of the mathematical questions associated with blackjack can be assessed with the mathematics we have already developed, without concern for the lack of independent events or the need to account for player decisions.

Example 6.2.1. Let's begin with a very simple question: What is the probability of being dealt a natural?

We consider only the player's hand, not the dealer's, and for simplicity, we will assume that we're dealing from a full single deck. Since a natural consists of a ten-count card and an ace, and since the order in which the cards arrive does not matter, we can consider each order separately. The probability of being dealt a ten-count card is $\frac{16}{52}$ and the probability of being dealt an ace is $\frac{4}{52}$. As cards are dealt, of course, these probabilities change, but since we're assuming that we start with a full deck, the only change is in the denominator, from 52 to 51. We then can compute

$$P(\text{Natural}) = \frac{16}{52} \cdot \frac{4}{51} + \frac{4}{52} \cdot \frac{16}{51} = \frac{32}{663} \approx .0482 \approx \frac{1}{21}$$

—so roughly 1 out of every 21 dealt hands will be a natural. In the calculation above, the first term in the sum covers the case where the ten-count card is dealt first, and the second covers the case where the ace is dealt first. Since these two events are mutually exclusive, the First Addition Rule may be used to add the probabilities.

If the hand is being dealt from the top of a six-deck shoe, a similar analysis gives

$$P(\text{Natural}) = \frac{96}{312} \cdot \frac{24}{311} + \frac{24}{312} \cdot \frac{96}{311} = \frac{4608}{97,032} \approx .0475,$$

slightly less than for a single-deck game.

That having been said, it must be noted that this probability is a long-term average value. If, in a one-deck game, all four aces are dealt on the first round (see Exercise 6.13), no further naturals are possible until the deck is reshuffled, but if the first round contains no aces or ten-count cards, the likelihood of a natural in the second round increases. Deck composition and the dependence of each hand on previously dealt hands are what make blackjack mathematics tricky and explain why card counting works to improve the player's chance of winning big bets.

An alternate way to compute long-term average probabilities is the *infinite deck approximation*. As the name implies, the method assumes that blackjack is being dealt from a shoe containing infinitely many decks of cards. This simplifying assumption means that the probability of drawing any particular card, or type of card, from the shoe is constant, and independent of how many cards have been drawn previously. Under this approximation, the probability of drawing a natural is

$$P(\text{Natural}) = \frac{16}{52} \cdot \frac{4}{52} + \frac{4}{52} \cdot \frac{16}{52} = \frac{8}{169} \approx .0473 \approx \frac{1}{21\frac{1}{8}},$$

a value which is quite close to the value of .0482 calculated above for a full single deck.

Example 6.2.2. Given a two-card blackjack hand, the dealer's upcard, and the specific game rules in force, it is possible through repeated simulation to obtain an accurate estimate of the probability p of winning the hand. At what probability is surrendering the correct choice?

If you surrender, your expectation on a $1 bet is a flat −$.50. If you choose instead to play out the hand, your expectation is

$$E = (1) \cdot p + (-1) \cdot (1 - p) = 2p - 1.$$

This will be greater than −$.50 if $p > .25$, so if you have at least a 25% chance of winning, surrendering is not the optimal play. Turning this around, if your hand has greater than a 75% chance of losing, you should surrender if the option is available.

The challenge here is that p is not a probability that is readily calculable from the three visible cards. Blackjack experts have, through the repeated simulation mentioned above, identified certain hands that should be surrendered, and these are listed in Table 6.2.

TABLE 6.2: Strategy for surrender [55]

If dealer shows	Surrender
Ace	10/6.
10	10/6, 10/5, 9/7, 9/6, 7/7

Note that the composition of the hand matters as well as the total: no soft hands should be surrendered, and while you should surrender 10/6 against a dealer ace, you should not surrender 9/7 or 8/8, which also add up to 16. A pair of 8s should be split rather than surrendered, even against a dealer ace.

Example 6.2.3. With an eye on the "all tie on dealer 22" rule in Blackjack Switch, it should be noted that 12 is the second most common two-card count, trailing only 20. This is in part because any first card can result in a two-card 12, whereas if your first card is a 9, there is no way to have a two-card total of 8. The probability of a two-card 12 in a single-deck game is

$$p = 2 \cdot \frac{4}{52} \cdot \frac{16}{51} + 2 \cdot \frac{4}{52} \cdot \frac{3}{51} + \frac{24}{52} \cdot \frac{4}{51} = \frac{248}{2652} \approx .0935,$$

or about once every 10.69 hands.

In this equation, the first term considers the case where the first card is a 2 or a ten-count card, the second covers the case where the dealt 12 is a pair of aces or a pair of 6s, and the third term collects all other cases leading to a hand of 12.

This calculation includes the hands 6-6 and ace-ace, which the player should often split rather than play out as 12s. When considering the possibility of a dealer drawing out to a total of 22, of course, these hands will not be split, for the dealer may not split pairs.

If we consider this question using the infinite deck approximation, then the probability of a two-card 12 is

$$p = 2 \cdot \frac{4 \cdot 16}{52 \cdot 52} + 2 \cdot \frac{4}{52} \cdot \frac{3}{52} + \frac{24}{52} \cdot \frac{4}{52} = \frac{31}{338} \approx .0917,$$

a respectably close approximation.

Example 6.2.4. How does the "different payoff on naturals" rule in the alternate blackjack games affect the player's income from naturals?

If we use the result of Example 6.2.1, we can conclude that once in every 21 hands, the player will receive a natural. If we're playing standard blackjack with a 3 to 2 payout, the extra money will be 50¢ per $1 bet per 21 hands, assuming *flat*, or constant, bets. A 6 to 5 payout gives the player an extra 20 per $1 bet per 21 hands, which is a 60% decrease from a 3 to 2 payoff gam

In either Super Fun 21 or Double Exposure 21, there is one type of natural that will pay an extra $1 per dollar bet. The probability of a natural in diamonds (as in Super Fun 21) is $2 \cdot \frac{1}{52} \frac{4}{51} = \frac{2}{663} \approx .003$, so the extra money accruing to the player is about $1 per $1 bet per 331.5 hands, or 3.17¢ $1 bet per 21 hands. This is less than the increased expectation of regular blackjack by a factor of about 16.

For Double Exposure 21, the probability of receiving the one blackjack that triggers a 2 to 1 payoff is $2 \cdot \frac{1}{52} \cdot \frac{1}{51} = \frac{1}{1326} \approx .00075$, so a player can expect to win an extra $1 per $1 bet per 1326 hands, or .016¢ per $1 bet per 21 hands, about $\frac{1}{63}$ of the payoff of standard 3 to 2 blackjack.

Blackjack Switch lacks even this single exception and so eliminates all of the advantage of a natural. While the option to switch cards means that Blackjack Switch leads to more two-card 21s than the other games, 21s arising from a card switch aren't considered natural 21s—not that that matters for the payoff.

We cannot fail to see the true impact of this rule change: since so much is being taken from the players with the different payoff on naturals, the casinos can afford to give quite a bit back with the new rules, especially considering that players might not adjust their playing strategy to account for these rules and thus hand over even more percentage points of advantage to the house. In light of this calculation, we are not surprised to see that switching from 3 to 2 to 6 to 5 payoffs on naturals, as many casinos in Las Vegas have done on their lower-limit or single-deck games, produces a larger HA.

That having been said, it is worth noting that rules can be changed. Some casinos have offered Double Exposure blackjack with a 3-2 payoff on naturals, and this payoff turns a house edge into a positive player expectation ranging from .4% to 2.1%, depending on the other rules in force (see Table 6.1) [99].

Example 6.2.5. On page 186, we introduced the Royal Match wager, which is an optional side bet on a player's first two cards whose resolution does not affect the play of the hand. Consider the Royal Match bet that pays $2\frac{1}{2}$ to 1 if the first two cards are suited (the "Easy Match"), 25 to 1 if they are the king and queen of the same suit (the "Royal Match"), and a progressive jackpot for the "Crown Treasure," which happens when both the player and dealer have a Royal Match.

Leaving out the Crown Treasure progressive bet for the moment, is this a bet worth making?

Once again, we begin by assuming a full single deck. For the Easy Match bet, the first card can be any card at all; we are simply looking for the probability that the second card is the same suit as the first.

$$P(\text{Easy Match}) = \left(\frac{52}{52}\right) \cdot \left(\frac{12}{51}\right) = \frac{12}{51} = \frac{4}{17} \approx .2353.$$

It should be noted that this calculation includes the probability of drawing a Royal Match. We will account for that when we develop the probability distribution.

For the Royal Match payoff, two things must happen:

1. The first card dealt to the player must be a king or queen.

2. The second card must be the rank not dealt in step 1, *of the same suit*

It follows that

$$P(\text{Royal Match}) = \frac{8}{52} \cdot \frac{1}{51} = \frac{2}{663} \approx .0030.$$

We find that the probability of hitting the Easy Match but *not* the Royal

Match is $.2353 - .0030 = .2323$. Defining X to be the return on a \$1 Royal Match bet gives the following probability distribution:

x	2.5	25	−1
$P(X = x)$	.2323	.0030	.7647

—so this is a bet you will lose approximately $\frac{3}{4}$ of the time.

For a \$1 bet on the Royal Match, we have

$$E = (2.5) \cdot .2323 + (25) \cdot .0030 + (-1) \cdot .7647 = -\$.10895,$$

and the HA is about 10.9%.

In the infinite deck approximation, we find that

$$P(\text{Easy Match}) = \left(\frac{52}{52}\right) \cdot \left(\frac{12}{52}\right) = \frac{12}{52} = \frac{3}{13} \approx .2308,$$

and that

$$P(\text{Royal Match}) = \frac{8}{52} \cdot \frac{1}{52} = \frac{1}{338} \approx .0030.$$

The difference between this last probability and the one calculated in the single-deck case is

$$\frac{2}{663} - \frac{1}{338} = \frac{1}{17,238} \approx 5.80 \times 10^{-5},$$

so little is lost in assuming an infinite shoe for the Royal Match payoff.

Example 6.2.6. If the Crown Treasure progressive bet is active, then the expected value of the Royal Match bet depends on the size of the progressive jackpot, and there will be an amount past which this bet favors the player. Let us denote the amount of the jackpot by A. The probability of winning the Crown Treasure bet is the probability that player and dealer both receive a Royal Match. For the dealer's hand, under the assumption that the player has a Royal Match, the probability of a second Royal Match is

$$P = \frac{6}{50} \cdot \frac{1}{49} = \frac{3}{1225} \approx .0024,$$

and so the probability of a Crown Treasure is

$$p = \frac{2}{663} \cdot \frac{3}{1225} = \frac{1}{135,362.5} \approx 7.388 \times 10^{-6}.$$

This probability, insignificant though it may be, must be subtracted from $P(\text{Royal Match})$ above in computing the new expected value. We have

$$E = (2.5) \cdot .2323 + (25) \cdot (.0030 - p) + (A) \cdot p + (-1) \cdot .7647$$
$$= -.10895 + p \cdot (A - 25).$$

If we set this last quantity equal to 0 and solve for A, we get

$$A = \frac{.10895}{p} + 25,$$

which yields A = 14,772.74438, so *if* the progressive jackpot for the Crown Treasure exceeds \$14,745.74, Royal Match is a bet that favors the player, since the expectation above will then be greater than zero.

Example 6.2.7. A challenge in developing blackjack side bets is separating the side bet from the play of the main game to minimize confusion, and so many such bets are resolved before players begin drawing cards. The *Lucky Lucky* side bet offers a range of payoffs based on the player's cards combined with the dealer's upcard, and pays off on any three-card total of 19 through 21, with a top prize of 200 to 1 if the three cards are all 7s of the same suit.

The probability of this last payoff depends on the number of decks in play (don't make this bet if it's offered at a double-deck game!). For a six-deck game, we are interested in the probability of drawing three cards and having them all be the same 7. Starting from a full shoe gives

$$p = \frac{4 \cdot {}_6C_3}{{}_{312}C_3} = \frac{80}{5,013,320} \approx 1.596 \times 10^{-5}.$$

200 to 1 scarcely seems adequate payoff for the incredibly long odds you've beaten in pulling this hand.

Example 6.2.8. Should you make the insurance bet?

If the dealer's upcard is an ace, you will be offered the opportunity to make what is effectively a side bet that the dealer has a natural. We assume that we're looking at the first hand of a new single deck, and that the only three cards known to you are the two in your hand and the dealer's ace. If you have no ten-count cards in your hand, the probability of a dealer natural is $\frac{16}{49}$. Since the insurance bet pays 2 to 1 and is limited to half your main bet, we have the following expected value for a \$1 main bet:

$$E = (1) \cdot \left(\frac{16}{49}\right) + (-.50) \cdot \left(\frac{33}{49}\right) = -\$.0102.$$

The HA on a 50¢ bet is thus 2.04%—not awful, but this assumes that you hold no tens. If you have one or two, the dealer has a smaller chance of completing a natural, and thus the HA of an insurance bet increases.

Holding one ten, we have

$$E = (1) \cdot \left(\frac{15}{49}\right) + (-.50) \cdot \left(\frac{34}{49}\right) = -\$.0408,$$

and the HA on the insurance bet is 9.16%.

If you hold two tens, then the expectation drops further, to

$$E = (1) \cdot \left(\frac{14}{49}\right) + (-.50) \cdot \left(\frac{35}{49}\right) = -\$.0714$$

—an HA of 14.28%.

Example 6.2.9. But what if you hold a natural yourself? Some casinos will offer even money on a player natural against a dealer ace—offering to settle the main bet as though an insurance bet had been made without the need for the player to make the bet. This even money option arises from the fact that if the dealer has a natural, your main bet pushes but a hypothetical insurance bet of half the main wager pays off at 2 to 1. The net effect is a profit of the amount originally wagered.

Psychologically, there is some appeal to this bet—at least you win even money on your one-hand-in-21 natural, even if the dealer ties you. You forfeit the extra 50% payoff at a 3 to 2 table, but that's equivalent to making an insurance bet and losing it. But what does the mathematics say?

As we might expect, the casinos aren't offering the even money option in an effort to be nice to players. There's something in it for them—let's do the math. There are 15 tens left in a 49-card deck. By taking even money, your return, regardless of the dealer's hand, is exactly \$1. If you decline even money, which is equivalent to not making an insurance bet, you win \$1.50 if the dealer does not have a natural and push if he does, so your expectation is

$$E = (1.50) \cdot \left(\frac{34}{49}\right) + (0) \cdot \left(\frac{15}{49}\right) \approx \$1.0408,$$

and declining even money is a better call by about 4.08%.

If, through bad fortune or poor judgment, you are playing blackjack at a casino that only pays 6 to 5 on naturals, then the expectation if you decline even money and the implied insurance bet is

$$E = (1.20) \cdot \left(\frac{34}{49}\right) + (0) \cdot \left(\frac{15}{49}\right) \approx \$.83,$$

so if the casino should offer even money ($E = \$1$) for your natural, take it. Of course, since casino management knows this, it is unlikely that you will even be offered even money on a natural if the casino pays only 6 to 5.

6.3 BASIC STRATEGY

Further calculation of blackjack probabilities will assume that the player is using the correct basic strategy, so a discussion of what that means is in order here.

The history of basic strategy begins with "The Optimum Strategy in Black-jack," a paper by Roger Baldwin, Wilbert Cantey, Herbert Maisel, and James McDermott that was published in the September 1956 issue of the *Journal of the American Statistical Association* [4]. The four authors were mathematicians serving in the Army at the Aberdeen Proving Grounds in Maryland, and their casual blackjack games inspired the foursome to analyze the game mathematically. This paper was the first to suggest that a set of rules for player action, which has now evolved into basic strategy, would lead to the best possible return for players. For the first time, the idea that the player's position in blackjack could be improved by skillful play was backed up with a rigorous mathematical argument. Among other revelations, this paper was the first to advocate that a pair of 8s should always be split and that standing on a 12 against a dealer 4, 5, or 6 is mathematically the best play.

Given the player's cards and the dealer's upcard, basic strategy gives the best choice for the player: when to hit, stand, double down, or split pairs for the best long-term results. The key word here is *long-term*: basic strategy, for example, tells a player to hit rather than stand when holding 15 against a dealer 8. Note that the only way a player 15 can win is if the dealer busts, since a completed dealer hand will always be 17 or greater. While hitting a 15 will certainly bust some hands—more often than not, in fact—computer simulation indicates that this action is best for the player, in the sense that, over a lifetime of gambling, you will win more (or lose less, which is effectively the same thing) by hitting and taking the chance of busting than by standing and hoping the dealer will bust.

The probabilities derived in this manner which give rise to basic strategy are often experimental probabilities, this being one case where it's easier to generate large numbers of simulated hands and draw conclusions from data than to account theoretically for the effects of player decisions on game outcomes.

Example 6.3.1. What does this mean mathematically? If the dealer's upcard is an 8, the probability distribution for the final value of the hand is (from [28]):

x	17	18	19	20	21	> 21
$P(x)$	.131	.363	.129	.068	.070	.239

so by standing on 15, you have only a 23.9% chance of winning; you will lose roughly three of every four hands. By hitting your 15, you take the chance of busting ($P(\text{Bust}) \approx \frac{7}{13} \approx .538$), but you also have about a 46% chance of improving your hand and possibly beating a nonbusting dealer hand. Your actual chance of winning if you hit is not quite 46%, but it's better than 23.9%, which is why basic strategy directs a player to hit 15 against an 8.

The goal of basic strategy is simple: to get the most money in play, via double downs and splitting pairs, when conditions favor the player, and to minimize exposure by forgoing these options when conditions are less favorable. If the dealer's upcard indicates that his hand is likely to bust—say, a 4, 5, or 6—then we double down and split pairs more frequently, and do not risk busting ourselves. When the hand is strong—a ten or ace showing, for example—then we take more risks as we try to compete against a hand that is more likely to be close to 21, and we don't invest additional money under the less favorable circumstances.

Table 6.3 contains the complete basic strategy for a multideck game. A simplified strategy, suitable for beginners, can be stated in a few straightforward rules:

1. **Always split 8s and aces.** Sixteen is a very weak hand, and turning it into two hands starting at eight gives you a fighting chance against the dealer. You may not win both hands, but your chance of breaking even, by winning one of the two new hands, increases. A player hand starting with an ace has a 52% edge over the casino; splitting a pair of aces gives you two such hands.

2. **Never split 4s, 5s, or 10s.** Eight may not be the best starting hand, but two hands starting with a 4 is a weaker situation. Splitting 5s means breaking up a hand of 10, which is a good total to draw to. Indeed, most of the time, a basic strategy player should double down on 5-5. Splitting 10s can be tempting, since starting a hand with a 10 gives you a 13% advantage over the house, but it means turning your back on a v strong hand of 20. In addition to being an unwise play, splitting 10s will almost certainly increase casino scrutiny and get you labeled as either a bad player or a card counter.

3. **Always double down on 11, and double down on 10 except against a 10 or ace.** In the infinite-deck approximation, holding an 11 gives you a 4/13 chance of drawing one card and getting a unbeatable 21. If you hit a 10 with one card, you have that same 4/13 chance drawing a 10 and reaching 20, and if you draw an ace ($p = 1/13$), you have a 21. In either case, you have a strong hand, and it's time to get more money in play.

4. **Hit any soft hand with a total less than 19.** A soft hand gives you two chances at improving your holdings. If drawing a card puts your total over 21, revalue the ace as 1 and follow the basic or simplified strategy for your new hand.

5. **If the upcard is 2 to 6, stand on any hard count of 12 or higher.** In short, don't risk busting when facing a weak dealer upcard.

6. **Against an upcard of 7 or better, stand only if your hand is 17 or higher.** The dealer is now more likely to have a strong hand, and you'll have to risk busting to improve your hand and have the best chance of winning.

TABLE 6.3: Basic strategy for six-deck blackjack [28]

Player's Hand		Dealer's upcard									
		2	3	4	5	6	7	8	9	T	A
	17+	S	S	S	S	S	S	S	S	S	S
	16	S	S	S	S	S	H	H	H	H	H
	15	S	S	S	S	S	H	H	H	H	H
	14	S	S	S	S	S	H	H	H	H	H
No pair or ace	13	S	S	S	S	S	H	H	H	H	H
	12	H	H	S	S	S	H	H	H	H	H
	11	D	D	D	D	D	D	D	D	D	H
	10	D	D	D	D	D	D	D	D	H	H
	9	H	D	D	D	D	D	H	H	H	H
	8−	H	H	H	H	H	H	H	H	H	H
	A/9	S	S	S	S	S	S	S	S	S	S
	A/8	S	S	S	S	S	S	S	S	S	S
	A/7	S	DS	DS	DS	DS	S	S	H	H	H
	A/6	H	D	D	D	D	H	H	H	H	H
Soft hand, no pair	A/5	H	H	D	D	D	H	H	H	H	H
	A/4	H	H	D	D	D	H	H	H	H	H
	A/3	H	H	H	D	D	H	H	H	H	H
	A/2	H	H	H	D	D	H	H	H	H	H
	A/A	SP	SP	SP	SP	SP	SP	SP	SP	SP	SP
	T/T	S	S	S	S	S	S	S	S	S	S
	9/9	SP	SP	SP	SP	SP	S	SP	SP	S	S
	8/8	SP	SP	SP	SP	SP	SP	SP	SP	SP	SP
Pair	7/7	SP	SP	SP	SP	SP	SP	H	H	H	H
	6/6	H	SP	SP	SP	SP	H	H	H	H	H
	5/5	D	D	D	D	D	D	D	D	H	H
	4/4	H	H	H	H	H	H	H	H	H	H
	3/3	H	H	SP	SP	SP	SP	H	H	H	H
	2/2	H	H	SP	SP	SP	SP	H	H	H	H
Never take insurance.											

Key:
D: Double down
DS: Double down if permitted, otherwise stand
H: Hit
S: Stand
SP: Split

7. **Never take insurance.** Example 6.2.8 gives the mathematical reason for this.

There are some places where this simplified strategy conflicts with basic strategy, but novices may find these rules easier to remember at first. Another option is to bring a wallet-size basic strategy card to the table—many casino gift shops sell such cards. There are also many blackjack Web sites (blackjackinfo.com is one) that allow you to specify the parameters of the game and will then generate a basic strategy chart appropriate for that game.

Example 6.3.2. For an examination of how basic strategy affects the house advantage, let us assume that we are playing six-deck blackjack under standard Las Vegas Strip rules (page 185). Suppose in addition that you choose to ignore basic strategy and simply adopt a "mimic the dealer" strategy by playing exactly as the dealer must: forgoing double downs and splits (since these are not available to the dealer), hitting on all 16s, and standing whenever your hand is 17 or higher, regardless of what card the dealer is showing.

Casinos will welcome you; indeed, they may compete for your business. In so doing, you are handing the house approximately a 5.46% advantage, more than they would get if you played American roulette [23]. Blackjack is symmetric under this approach, since both sides are using the same strategy. Remembering that the HA in blackjack derives from the double bust, and that the dealer's required strategy results in a bust 28% of the time, the HA starts at $(.28) \cdot (.28) = .0784$. We must reduce this to account for the 3 to 2 payoff on naturals available to the player but not to the dealer: this reduction is approximately $.5 \cdot \frac{1}{21} = 2.38\%$, since the player profits an additional half a bet every 21 hands on average (see Example 6.2.1). Subtracting gives an approximate HA of 5.46%.

Using the recommendations called for in the basic strategy table gives a player the benefits in Table 6.4, which have been determined by repeated simulation [23, p. 19].

TABLE 6.4: Basic strategy advantages

Option	Edge
Proper standing	3.2%
Doubling down	1.6%
Proper pair splitting	.4%
Hitting soft 17 & 18	.3%

Adding these together shows that basic strategy can turn the HA from 5.5% to approximately 0%—a dead-even game. It is admittedly difficult to find a game played under these rules today; many casinos offering otherwise favorable games extract a price from players by paying only 6 to 5 on naturals.

This modification raises the HA by 1.39%, a change roughly equivalent to banning double downs. A 6 to 5 payoff on naturals is, by itself, good reason to seek out a different table, or a different casino entirely, for your blackjack action. Moreover, casinos that require that a dealer hit soft 17s are giving themselves a chance to turn a weak dealer hand into a stronger one—admittedly at some risk—and increasing their advantage by about .20% [28, p. 133].

Going in the other direction: From time to time, some casinos will offer a promotion or coupon where blackjacks pay 2 to 1. This adds 2.37% to the player's expectation in a six-deck game, turning a 1% HA into a 1.37% player edge. Benny Binion, longtime owner of the Horseshoe Casino (now Binion's) in downtown Las Vegas, used to offer this promotion at Christmas as a gift to his loyal players—but he limited his losses by restricting the bonus to bets of $5 or less and only allowing it on one hand per player at a time. If you were playing more than one hand at a time, only one hand qualified for the 2 to 1 payoff.

Of course, basic strategy errors do occur, whether because players find themselves in a rare situation that they have not memorized—for example, whether or not to split 9s against a 7 (you shouldn't, because your total of 18 will win if the dealer has a ten in the hole, which happens with probability $\frac{4}{13}$)—or because they are "feeling lucky" and acting on gut instinct rather than following basic strategy recommendations. Such errors, averaged across all players, are generally taken to add about .7% to the casino's advantage [102].

Variations

It should be noted here that blackjack variants—Double Exposure, Spanish 21, and others—require different strategies if the player is to play correctly and minimize the house advantage. For example, in Spanish 21, pursuing the bonuses for 21s with five or more cards sometimes calls for a player to hit hands with a number of small cards in hopes of catching the bonus. Blackjack Switch strategy includes guidelines for how and when to switch cards between hands, which must be mastered if a player is to minimize the HA, which is .58% in a six-deck game, and also modifies basic strategy slightly to accommodate the "tie on dealer 22" rule [69]. These latter modifications call for fewer double downs and splits, and are intended in part as protection against the high HA that "push on 22" affords the casino; this rule adds about 6.91% to the casino edge [68].

In Multiple Action 21, however, the correct basic strategy is that for standard blackjack. Multiple Action players must resist the temptation to play cautiously in an effort to avoid busting and immediately losing three bets. Rather than changing their style of play, skilled Multiple Action players account for the increased risk by betting slightly less on their one hand than they would bet on three separate hands. In [98], Stanford Wong reports that the optimal bet level for a Multiple Action player betting against two dealer hands is 70% of his or her standard single-hand bet, and the optimal level for three hands is 53% of that single-hand wager. In practice, a player who typically bets $100 per hand at a standard blackjack game should make two bets of $70 or three bets of $50 in Multiple Action. These figures were obtained, as is often the case in blackjack, by repeated simulation of many hands.

Example 6.3.3. The Las Vegas Sands and the Dunes casinos are no more, but if you're playing blackjack with a joker, as was pioneered at their gaming tables, what should you do if you're dealt a wild joker?

Some choices are obvious: if you're dealt a face card and a joker, make the joker an ace and your hand becomes 21. Similarly, a joker arriving with an ace should be counted as a 10. Other situations are far less clear: what should you do, for example, when you draw a joker as your third card to an initial hand consisting of a deuce and four?

There are no 14s or 15s in a standard deck, so you won't be reaching 20 or 21. The best strategy, then, is to position yourself to draw a good fourth card. This is best accomplished by counting the joker as a five, making your hand an 11. Your next card cannot bust your hand, and you stand a good chance of drawing a 10 and pulling a total of 21.

As a general rule, the following algorithm (from [98]) should guide your actions in assigning a value to a joker:

1. If possible, bring your hand to a total of 21.

2. If possible, bring your hand to a total of 20.

3. Either bring your hand to a total of 11 or call the joker an ace.

When a joker is dealt as one of the first two cards, the last decision depends on whether or not doubling down on jokered hands is permitted. If this is allowed, then 8+joker should be called an 11 against a dealer upcard of 2 to 6 or 10—you should then double down. Against an upcard of 7, 8, 9, or ace, call your hand 19 and stand. Other joker hands should be brought to 11 and then doubled unless the dealer's upcard is an ace.

If the casino forbids doubling down on joker hands, then the third step depends on the dealer's upcard.

- On 8+joker, call the joker a 3 and hit against a 10; otherwise call the joker an ace to make the hand 19 and stand.

- On 7+joker, call the joker an ace and stand against a 7; otherwise call it a 4 and hit on your 11.

- For all other two-card hands with a joker, make the total 11 and hit.

With this strategy, the player's edge increases by 4.2% per joker per deck [98]. Adding three jokers to a six-deck shoe represents half a joker per deck and so increases the player's edge by 2.1%—enough to tip the overall advantage to the player if standard Strip rules are used otherwise.

Example 6.3.4. The "5-Card Charlie" and "6-Card Charlie" are perhaps more common in casual home blackjack play than in the casino game. This rule variation, a version of which is incorporated into Super Fun 21, states that a five- or six-card player hand totaling 21 or less is an automatic winner regardless of the dealer's hand.

To take full advantage of this rule change necessitates some small changes to basic strategy—to wit: occasionally drawing on some four- or five-card totals in hopes of catching a small fifth or sixth card and an immediate winning hand. For example, if you hold a soft 21 in five cards, such as A-2-2-2-4, drawing a sixth card puts you at no risk of busting and guarantees you a win, even if the dealer later draws out to the 21 that would tie you if you stood on your 21.

"Charlies" are perhaps best analyzed, once again, through simulation. Stanford Wong has described the necessary modifications to basic strategy if the 5-card Charlie payoff is active; this rule increases the player edge by 1.46%, *provided* that the player uses the correct modified basic strategy. Under the assumption that a 5-card Charlie pays even money, those modifications are detailed in Table 6.5. In some cases, these strategy changes call for forgoing doubling down on a soft hand so that additional cards may be drawn if the bonus is reachable.

TABLE 6.5: Basic strategy modifications for 5-card Charlie blackjack games [98]

4-card hands	Against a dealer
Hit soft 16	4
Hit soft 15	5
Hit soft 14	5 or 6
Hit soft 13	6
Hit hard 17	8, 9, 10, or ace
Hit hard 16	2 or 3
Hit hard 15	2 through 6
Hit hard 14	2 through 6
Hit hard 13	3 through 6
3-card hands	**Against a dealer**
Hit hard 13	2
Hit hard 12	4, 5, or 6
2-card hands	**Against a dealer**
Split aces	10 or ace
Split 3s	4, 5, or 7

A 6-card Charlie rule lowers the HA by .16%. Table 6.6 contains the recommended strategy changes for a game where any six-card hand is an automatic winner.

TABLE 6.6: Basic strategy modifications for 6-card Charlie blackjack games [23]

5-card hands	Against a dealer
Hit soft hands less than 20	Anything but 7 or 8
Hit hard 15 and less	2
Hit hard 14 and less	3 or 4
Hit hard 13 and less	5 or 6

6.4 CARD COUNTING

As we have repeatedly mentioned throughout this chapter, blackjack is different from other casino games in that successive trials are not independent. Each hand in blackjack depends in a meaningful way on the cards that have been dealt since the last shuffle, and as a result, probability calculations for blackjack must take this into account. The infinite deck approximation described in Section 6.2 assumes independence in its calculations and sacrifices a small amount of accuracy for ease of computation. Since we are looking specifically at how previous cards affect probabilities, we must discard the assumption of independence and thus the infinite deck simplification.

Mathematician Edward O. Thorp raised the profile of basic strategy as devised by Baldwin et al. and added the new twist of card counting with his 1962 book *Beat the Dealer* [87], in which he presented a mathematical argument describing how the composition of the undealt cards in a deck affected the probabilities of hands yet to come. In the ensuing years, blackjack players and mathematicians alike worked to determine an optimal strategy for keeping track of unplayed cards and using that information to adjust betting and playing strategies—and casino officials sought to counteract this new player advantage by changing the rules of the game and the procedures for its operation.

There is nothing mathematically complicated about card counting; all that is required for the most commonly used counting system is the ability to add and subtract 1, and to divide. (To do so under casino conditions is, as portrayed in the movie *21*, considerably more challenging.) It is not necessary, nor is it desirable, to try and track every card in a deck or shoe; all that most card-counting systems do is give the player a sense of the balance between high and low cards in the remaining undealt portion of the deck. This information, combined with the ability to adjust bet sizes and the willingness to learn exceptions to basic strategy, makes card counting a tool that can give a blackjack player an advantage over the casino.

Casino managers, of course, know this, and so the years since the publication of Thorp's book have seen a collection of changes to blackjack as casinos try to stop, or at least discourage, card counters. These changes include the introduction of more decks into the game, decreased penetration before reshuffling, and more conservative rules, as well as increased scrutiny of player activity, especially bet fluctuation.

Thorp's Five-Count

The simplest card-counting system described in *Beat The Dealer* involves tracking only the 5s. This is best suited for single- and double-deck games; when the book was first published, shoes containing four, six, or eight decks were nowhere near as common as they would become. Since the event that triggers a change in betting and playing strategy is the removal of all the 5s, this is far less viable in shoe games, where the likelihood of all the 5s being dealt out before the cards are shuffled is very small.

Why the focus on the 5s? Since the dealer must draw when holding 16, the absence of 5s renders the deck less favorable to the house. Thorp's calculations measured the effect on the house advantage of the removal of all the cards of each rank from a single deck, and his results are shown in Table 6.7. It can be seen that the biggest boost to the player comes when all of the 5s are removed, whereas removal of all of the aces or 9s increases the house edge.

TABLE 6.7: Effect of card removal on blackjack house edge [87]

Card Rank	A	2	3	4	5
% HA Change	2.42	−1.75	−2.14	−2.64	−3.58
Card Rank	6	7	8	9	10
% HA Change	−2.40	−2.05	−0.43	0.41	−1.62

It follows that, if all of the 5s have been dealt, and there are sufficient cards remaining in the deck to complete the next hand, the player has a considerable edge, and a larger bet is therefore called for. Some small changes in basic strategy are appropriate in a game with no 5s remaining; among these is that a pair of 10s should be split against a dealer's upcard of 6 [87].

The High-Low Count

A simple but powerful counting system, and arguably the most popular, is the *high-low count*. The idea behind the high-low count is that a deck with a surplus of high cards favors the player, in the sense that the probability of naturals and of drawing a high card on a double down both increase. Of course, the probability of the dealer pulling a natural also increases, but since player naturals are paid 3 to 2 while dealer naturals lose only the amount bet, this is still a benefit for the player.

Thorp's original high-low card-counting system simply tracked the ratio of non-10s to 10s in a deck, a ratio that starts at 36/16 in a full deck [87]. This was soon eclipsed by more sophisticated counting methods. In the high-low counting system, the cards are assigned the following point values:

Cards	Value
Low: 2, 3, 4, 5, 6	+1
Neutral: 7, 8, 9	0
High: 10, J, Q, K, A	−1

A complete deck, or shoe of several decks, has initial count 0, since there are exactly as many high cards as low cards in a deck. Cards are counted as they appear: add 1 to the count for every low card dealt and subtract 1 for every high card. The total is called the *running count*, RC.

Example 6.4.1. At the start of a four-player single-deck game, suppose that the cards dealt are J/9, K/5, K/J, and 2/Q, with a 4 for the dealer's upcard. The running count is $(-1+0) + (-1+1) + (-1-1) + (1-1) + 1 = -2$.

For playing purposes, the running count must be converted into the *true count*, TC. This is done by dividing the RC by the number of decks remaining to be dealt. This divisor is estimated from the progress of the hand; a look at the discard rack on the table can help with the estimation. This modification gives a per-deck accounting of the excess or deficiency of high cards. A running count of +4, signifying an excess of four high cards over low cards, is a lot more meaningful if only one deck remains than if three are left in the shoe.

Example 6.4.2. Completing the hand above: Since the dealer's upcard is a 4, basic strategy dictates that all four players, who hold 19, 15, 20, and 12, stand. No further cards were drawn and the count remained at −2. The dealer's hole card was a 7 (RC still −2) and he drew a 10 (RC now −3) for a total of 21, beating all four players. Eleven cards have been dealt, leaving roughly 3/4 of a deck—in general, the number of decks remaining is estimated to the nearest half- or quarter-deck. The TC is then $-3 \div (3/4) = -4$, meaning that low cards currently exceed high cards, making the game less favorable to players. Consequently, they should bet no more than the table minimum.

There are two uses for the TC. The first is to indicate a point when the player's initial bet should be increased because the cards are favorable and a good hand is more likely, or conversely (as in the example above), when the cards favor the casino and bets should be kept as small as possible. To be precise, every increase of 1 in the true count translates into approximately a .5% advantage for the player [91]. Since the game starts out with a house edge of about .5%, a count of +1 makes the game essentially even, and if the true count is +4, players have a 1.5% edge over the casino.

The second purpose of the TC is to highlight times in the course of a deck or shoe when the composition of the remaining cards is such that certain deviations from basic strategy are, probabilistically, more favorable to the player. Once again, these true counts, called *index numbers*, have been determined by computer simulation.

For example, consider the insurance bet. Since you're basically making a side bet that the dealer has a ten in the hole, a positive count indicating more high cards might indicate that that bet is worth making. Since insurance pays 2 to 1, this bet has a player advantage (independent of the player's hand) if the ratio of 10s to non-10s exceeds 1:2.

Example 6.4.3. Suppose that on the first round of a one-deck game, 11 cards were dealt: 2 ten-count cards and 9 lower cards. On the second round, the dealer shows an ace and your hand is a pair of 5s. The composition of the deck is now $16 - 2 = 14$ ten-count cards and $36 - 12 = 24$ nontens. Since the ratio of tens to nontens is $14/24 > 1/2$, an insurance bet has an expectation in your favor. Specifically, the expected value of a 50¢ insurance bet is

$$E = (1) \cdot \frac{14}{24} + (-.50) \cdot \frac{10}{24} = \$\frac{9}{24} > 0,$$

so you should make the insurance bet.

This does not account for the fact that if the insurance bet wins, your original $1 bet loses unless you also have a natural, and a careful analysis of game options will have to take that into account.

Translating this into the world of card-counting, this means that if the TC is +3 or greater, the insurance bet favors the player. This index number takes into account the fact that aces as well as tens count −1 in the high-low count, hence the count by itself is not a perfect indicator of the ratio of tens to nontens remaining in the shoe. At the other end of the spectrum, a highly negative true count, indicating an abundance of low cards over high ones, may mean that hitting certain hard hands despite the advice of basic strategy—for example, hitting a hard 12 against a 5 at a TC of −2—is called for. Eighteen deviations from basic strategy for a four-deck game where the dealer stands on all 17s—dubbed the "Illustrious 18" by blackjack experts—are listed in Table 6.8 in decreasing order of their contribution to the player's edge.

When playing a game where surrender is allowed, judicious use of this option in conjunction with the true count can further add to a player's edge. In addition to the recommended surrenders in Table 6.2, the "Fab 4" surrender decisions listed in Table 6.9, again in order of the advantage they carry, are profitable at or above the indicated count [61].

It is important to understand the inherent risk here: Part of the appeal to casinos of offering the surrender option is that many players will not use it even when it's the "correct" decision. A blackjack player who surrenders correctly may be a card counter, and permitting this game choice may be part of a casino's strategy for identifying and barring counters.

A closer analysis of the effect on the player's advantage of removing individual cards from the deck can be found in [23], where theoretical and experimental evidence were combined in the following model: For each of the ten card ranks, define the following variables:

- x_1 = The number of stiff hands (12 to 16) which a particular card wi improve to 17 or higher.

TABLE 6.8: Blackjack's Illustrious 18 [61]

Strategy deviation	If the high-low true count is
Take insurance	$\geq +3$
Stand on 16 vs. 10	≥ 0
Stand on 15 vs. 10	$\geq +4$
Split 10s vs. 5	$\geq +5$
Split 10s vs. 6	$\geq +4$
Double 10 vs. 10	$\geq +4$
Stand on 12 vs. 3	$\geq +2$
Stand on 12 vs. 2	$\geq +3$
Double 11 vs. ace	$\geq +1$
Double 9 vs. 2	$\geq +1$
Double 10 vs. ace	$\geq +4$
Double 9 vs. 7	$\geq +3$
Stand on 16 vs. 9	$\geq +5$
Hit 13 vs. 2	≤ -1
Hit 12 vs. 4	≤ 0
Hit 12 vs. 5	≤ -2
Hit 12 vs. 6	≤ -1
Hit 13 vs. 3	≤ -2

TABLE 6.9: Blackjack's Fab 4 [61]

Surrender	If the high-low true count is
14 vs. 4	$\geq +3$
15 vs. 10	≥ 0
15 vs. 9	$\geq +2$
15 vs. ace	$\geq +1$

- $x_2 = $ The number of stiff hands which a particular card will bust.

- $x_3 = 1$ if the card is a ten-count card, 4 if the card is an ace, and 0 otherwise.

For example, if the card in question is a 3, we have $x_1 = 3$, $x_2 = 0$, $x_3 = 0$. Simulation of many hands indicated that removing an ace, 9, or 10 decreased the player's edge, removing an 8 had no effect, and removal of any other card increased the player's advantage. The cumulative effect of card removal, denoted by y, was found to be well-modeled by the formula

$$y = .14 \cdot x_1 - .07 \cdot x_2 - .18 \cdot x_3.$$

The values predicted by this formula are all within .1 of the actual effect determined by simulation, and in most cases are far more accurate than that.

The K-O Count

After the introduction of the high-low count, a number of blackjack experts set out to improve on it, whether through more sophisticated counting schemes that purported to provide more complete information or simpler schemes that were said to be easier to implement. The K-O count, named for its inventors Ken Fuchs and Olaf Vancura, is described as "the easiest card-counting system ever devised" [94]. An advantage of the K-O system is that it immediately produces the true count as play elapses, without the need to estimate the number of decks remaining or to divide by that estimate.

K-O is an *unbalanced* counting system, meaning that the total count of a complete deck or shoe is not 0. The point system for individual cards in the K-O system is

Cards	Value
Low: 2, 3, 4, 5, 6, **7**	+1
Neutral: 8, 9	0
High: 10, J, Q, K, A	−1

This differs from the high-low count only in that 7s are classified as low cards, with a value of +1, rather than as neutrals. With this modification, the value of a full deck is +4, and the value of an n-deck shoe is $+4n$.

The necessary twist in the K-O count is that the count at the start of a shoe starts at $4 - 4n$, where n is the number of decks in use. If you're playing a single-deck game, this means that you start at 0, as with the high-low count. In a six-deck shoe game, the count starts at −20. From this starting number, cards are counted as they appear, as in any other counting system, and the running count is updated.

Example 6.4.4. In a six-deck game with five players, the following hands are dealt:

$$A\spadesuit 10\clubsuit, 7\heartsuit 5\spadesuit, J\clubsuit 5\diamondsuit, 9\heartsuit 7\spadesuit, 6\heartsuit 5\diamondsuit.$$

The dealer's upcard is an $A\spadesuit$. The K-O count, working from left to right, is

$$(-20) - 1 - 1 + 1 + 1 - 1 + 1 + 0 + 1 + 1 + 1 - 1 = -18.$$

The high-low count for this hand would be 0, since the count would start at 0 and the two 7s would be counted at 0 instead of +1.

Making this counting system easy is this: The true count for the K-O count is the same as the running count. No division is necessary, and there is no need to estimate the number of decks remaining. Decisions relative to changing deck composition are made by referring to the *key count* for the number of decks in play. This is a fixed number that, unlike the running count, does not fluctuate as cards are dealt. The key count is the number at which a K-O counter has an advantage over the casino [94]:

Decks	Key count
1	+2
2	+1
4	−1
6	−4
8	−6

Example 6.4.5. As a comparison between the K-O count and the high-low count, consider a six-deck game. If the K-O count, which starts at −20, has risen to −4 after, say, two decks have been dealt, this means that 16 more low cards than high cards have been dealt. Two decks contain eight 7s, counted as low cards in K-O, so the average corresponding high-low running count would be +8. Dividing by the four decks remaining gives a high-low true count of +2, and this indicates a .5% player edge using basic strategy alone.

Betting in light of the K-O count is simple: For K-O novices, Vancura and Fuchs recommend a two-level wagering structure: make a "small" bet when the count is less than the key count and a "large" bet when the count exceeds this number. If the large bet is five times the small bet, a K-O counter can achieve an edge over the house ranging from .16% in an eight-deck game to .88% at a single-deck table [94, p. 77]. In addition to being simple, this betting strategy has the advantage of evading casino scrutiny, since the bet is not moving with the count, as is often done with other counting systems that call for larger and larger bets as the count grows more favorable.

For a more advanced version of K-O, there are also changes in basic strategy associated with this count. In this "K-O Preferred" strategy, insurance becomes a good bet whenever the count—with any number of decks—is at least +3. Using this modification alone adds from .16% in an eight-deck game to 1.06% at single-deck to the edges quoted above for a 1- to 5-unit bet spread [94].

Example 6.4.6. +3 is, of course, the same value given for the high-low count as the benchmark for insurance bets. Consider a two-deck game in which approximately 26 cards, or half a deck, have been dealt out. If the high-low true count is +3, this means that the running count is approximately +5, indicating an excess of five low cards over high cards dealt to the table. If 26 cards have been dealt, we would expect that two 7s have appeared on average, and so the K-O count would be -4 (starting value) + 5 (for the low cards from 2 to 6) + 2 (counting 7s as low cards), or +3 again.

Under either counting system, an insurance bet would be called for if the dealer's upcard is an ace.

Since the K-O count does not divide by the number of remaining decks, additional modifications to basic strategy that are called for by the K-O count are based on index numbers. The size of the shoe is accommodated by the fact that, in some cases, these indices depend on how many decks are in play. The recommended modifications are listed in Table 6.10.

TABLE 6.10: Modifications to basic strategy based on the K-O count [94]

Decks					
1	**2**	**4**	**6**	**8**	**Strategy deviation**
+3	+3	+3	+3	+3	Take insurance
+4	+4	+4	+4	+4	Stand on 16 vs. 9
					Stand on 15 vs. 10
					Stand on 12 vs. 2
					Stand on 12 vs. 3
					Double 11 vs. ace
					Double 10 vs. 10
					Double 10 vs. ace
					Double 9 vs. 2
					Double 9 vs. 7
					Double 8 vs. 5
					Double 8 vs. 6
+2	+1	−1	−4	−6	Stand on 16 vs. 10
0	−4	n/a	n/a	n/a	Hit 13 vs. 2
					Hit 13 vs. 3
					Hit 12 vs. 4
					Hit 12 vs. 5
					Hit 12 vs. 6

These changes in playing strategy should be used as follows:

- For the "stand" and "double" decisions, deviate from basic strategy whenever the K-O count equals or exceeds the index number in t table. These decisions are correct if the shoe holds a surplus of high cards, as indicated by the count.

- For the "hit" decisions, deviate from basic strategy when the K-O count is *less than* the index number in the table. These decisions are riskier against a shoe containing a surplus of high cards, and so should only be used when the remainder of the shoe is relatively rich in low cards.

Note that the strategy associated with the K-O count includes *no* changes to basic strategy for splitting pairs or handling soft hands. Additionally, certain strategy changes are not recommended in a game with more than two decks; these are identified by "n/a" in the table.

Casino Pushback

It must be stressed at the outset here that *card counting is not illegal* Card counters are simply using their brains to keep track of what's going on at the table, and it is difficult to imagine the magnitude of the fallout from a law forbidding that. However, while they are violating no laws, that does not give them free rein to practice their craft. Casinos, being private businesses, are free to restrict or forbid access to anyone they choose, and so casinos in Nevada may, if they wish, eject suspected card counters.

New Jersey law, by contrast, forbids casinos from barring card counters; Atlantic City casinos deal with counters by restricting the amount they are allowed to bet or by shuffling after every hand, and so eliminating any advantage that might be gained from counting cards. Bob Stupak followed a similar policy at Vegas World, where he was known to issue identified card counters a card stating that they were welcome to play blackjack, but that their wagers were to be restricted to no more than 7 units [79]. This meant that, for example, at a $5 table, known counters could not bet more than $

Casino reactions to card counting seek an elusive balance between the fact that successful card counters have the potential to win a lot more money than the house advantage says that they should and the equally important fact that far more people think they can count cards successfully than actually can, which leads to increased traffic at blackjack tables and, ultimately, the likelihood of increased casino income from the tables.

Since the first of these facts is easier for casinos to address, we begin by describing some casino countermeasures. One of the first things that casino managers did after *Beat the Dealer* was published was to tighten up the rules of the game and so eliminate the advantages of card counting. On April 1, 1964, the Las Vegas Resort Hotel Association announced that henceforth, double downs would be allowed only on hands counting 10 or 11 and that blackjack players were forbidden to split aces [88, p. 128]. Table 6.1 shows the effect of these changes: the HA with these new rules went up by .44%. Blackjack players—counters and noncounters alike—stayed away from the games with new unfavorable rules, and they were reversed within three weeks.

After that reversal, casinos started looking for other game modifications that might restore their advantage and started moving from single-deck hand-dealt games to multiple-deck shoe games. Adding decks, in addition to increasing casino profits because more time is spent dealing the cards than shuffling them, increases the HA. Table 6.1 (page 185) gives the effect on the HA of using more than one deck. Note that the increase in the house advantage levels off as more decks are added—the casino derives no significant game-related advantage in going past eight decks, and the increased height of the stack of cards may make it unwieldy for a dealer to handle.

Card counters then responded by developing new guidelines for counting into multideck shoes, which led to the running count and true count, and so the arms race of sorts between players and casinos was on. As casinos found a new rule or game procedure that would improve their position, the card-counting community developed a modification to their tactics that would combat it, and vice versa.

Since one important part of card counting is bet fluctuation—you bet more when the count is highly positive to take advantage of favorable conditions—

casinos began barring players whose bets went up and down too widely over a period of observation, or using *preferential shuffling*: shuffling the cards as soon as a player increased their bet. This worked both ways, of course—a counter could increase his or her bet during a highly *negative* deck to encourage the dealer to shuffle early and give up the house's edge. At many casinos, dealers are not part of the defense against card counters, as they have enough to do just running the game properly. As a result, the dealer might well have been unaware of the count and simply reacted to the jump in the player's wager.

Even nonpreferential shuffling sometimes works in a casino's favor. In a single-deck game with five players against the dealer, one can safely expect two full rounds to be dealt, since an average blackjack hand uses 2.7 cards. With six hands per round, we would expect 33 cards to be dealt in the first two rounds, leaving only 19 for a third round. Many dealers will shuffle the cards when less than half a deck remains, so two rounds is about all one would expect—unless those first two rounds use an unusually small number of cards. When would that happen? Precisely when most players stand on their initial two-card totals, and this frequently happens when the player's hands are rich in high cards, leaving a partial deck with a highly negative count, favoring the casino, on the third round. This benefit accrues to the casino whether or not the players are counting cards.

A significant player reaction to casino action against bet variation was the idea of team play, where certain team members would play and count while flat-betting, then when the count was high, call in a "big player" who came in, made a few large bets, and left when the count dropped. Since the big player simply made large bets at advantageous tables, there was no bet variation to arouse suspicion—to all appearances, he or she was just a lucky high roller. Some of the most famous blackjack teams were based at MIT; their exploits were chronicled by Ben Mezrich [44, 45] and portrayed in the movie *21*.

Casinos countered team play by forbidding mid-shoe entry, requiring players to wait for the shuffle before joining a game and thus blocking "big players" from jumping into favorable shoes. This also protected the casino against *wonging*: the practice where an individual player counts down a deck or shoe while standing behind the table instead of wagering, then settles in to play only if the count grows favorable.

At the same time, the publication of the "Illustrious 18" and other recommended deviations from basic strategy led casinos to watch players somewhat more closely for such deviations. A player who splits tens when the deck is highly positive (part of the Illustrious 18 if the dealer's upcard is a 5 or 6), for example, is almost certain to attract scrutiny from casino personnel, for this is a move that is almost never made by noncounters. A player who splits 10s is either very inept—and thus a valued customer—or a card counter. Ongoing observation of the player's actions relative to the count—and players should take it for granted that *someone* on the casino security staff knows how to count cards—will easily determine which. The casino can then act accordingly: encouraging the inept player with complimentary meals, hotel rooms, or other inducements, and barring the counter or initiating countermeasures to make the game less attractive.

In the early 1970s, some card counters turned to the new technology of home-built portable computers to automate and expedite the counting process. Technical restrictions hampered many of these devices, which had to be small enough to be undetectably concealed on a player's person. Some were controlled through switches operated by a player's toes, and these were frequently unreliable. Additionally, some means had to be found to signal the correct playing decision to the player, and this posed another significant challenge. Radio transmitters ran the risk of detection when they interfered with casino security systems, and smaller devices that connected to fake digital watches or LEDs imbedded in eyeglass frames were tricky to conceal. This age of innovation emphatically ended in Nevada on July 1, 1985, when portable computers were banned in casinos. Violators of this law are subject to fines of up to $10,000 plus 1 to 10 years in jail [93]. New Jersey followed with its own law, and as new gaming jurisdictions came online, they swiftly moved to ban card-counting computers.

This issue resurfaced beginning in 2009, when the introduction of Apple's iPhone led several developers to program and market card-counting applications. Some of these apps were designed to operate even when the screen was turned off, and so avoid the giveaway of an illuminated screen. Casino operators were warned about the apps, and the Nevada Gaming Control Board swiftly forbade their use in a casino. Card counting may be legal, though not exactly welcome, but using an electronic device to count cards is emphatically not.

One more recent device introduced by casinos to counteract card counting is the *continuous shuffling machine* (CSM). A CSM is a machine that collects the discards from each hand and immediately re-inserts them into the packet of cards, thus effectively dealing every hand from a fresh shoe of four to eight decks and eliminating all of the advantages of card counting. As when casinos addressed the advent of card counting by changing the rules to restrict player actions in the 1960s, CSMs as a casino response to advantage play proved to be something of an overreach. Part of the continuing appeal of blackjack is the fact—and it is a fact—that it can be beaten with sufficient work. Continuous shuffling machines, by completely negating the advantages to be gained from counting cards, demolish this selling point. While CSMs are still to be found on some casino floors, serious gamblers avoided them so assiduously that blackjack income dropped off. In a competitive market such as Las Vegas or Reno, gamblers had options, and exercised them. In a market with a small number of casinos such as Osceola, Iowa (home to only one casino, the Lakeside), blackjack players must take the game as offered and so have little recourse to resist any anti-counting strategies. Of course, smaller markets such as Osceola are less susceptible to major attacks by card counters.

Card counters are obviously disadvantaged by CSMs, but do they affect ordinary basic strategy players? The play of an individual hand is not affected for noncounters; the effect kicks in to their detriment when we consider the increased game pace that CSMs allow. In light of the fact that no playing time is lost to shuffling the cards, a CSM allows the dealer to deal more hands per hour than an ordinary shoe, and this exposes the players to more hands where the casino's advantage works against them.

It should be noted that there's a difference between a CSM and an automatic shuffler. Many casinos have turned to shuffling machines in order to reduce the time spent not dealing blackjack, but only if the cards are continuously replaced in the shoe are you dealing with a CSM. If the used cards are collected in a tray for multiple hands before being inserted into the machine, and if there's a pause in the game while the cards are machine shuffled, then the only disadvantage to players—counters or not—comes from the aforementioned increased number of hands dealt per hour.

With regard to the second fact described above, the tendency of many casino patrons to overestimate their ability to count cards, casino management simply needs to find a way to welcome these players, who pose no real threat to their blackjack tables, for low-level gambling that won't cost the casino too much in the event of a short-term run of player luck. This process includes the realistic assessment of anti-counter methods for their effect on the casino's overall financial health. Casino management expert Bill Zender has advised casinos against overreaction to the supposed threat posed by counters, claiming that excessive countermeasures had the unintended effect of decreasing the casino's profits by wasting time trying to respond to an overstated threat. Zender's assertion is that increasing the number of blackjack hands dealt per hour is key to improving a casino's bottom line, and such tactics as excessively complicated shuffles (to deter "shuffle trackers," an advanced type of player who tries to track the movement of groups of cards through the shuffle and bet appropriately when the same cards come up in the next round [45]), barring mid-shoe entry, and dealing less than 80% into a shoe (75% in hand-held single- or double-deck games) work against maximum dealer efficiency [102].

As casino manager at the Aladdin Casino in Las Vegas (now Planet Hollywood) during the mid-1990s, Zender instituted a number of policies that, taken together, had the effect of producing one of the best blackjack games offered on the Strip at the time [43]. The Aladdin offered the following rules on its blackjack tables [102, p. 17–22]:

- *Single-deck games replaced by double-deck games.* Zender's contention was that players don't readily distinguish between the two games, classifying them all as "hand-held." Since single-deck blackjack is approximately an even game for players using perfect basic strategy, and is specifically sought out by expert players, this move to two decks benefits the casino without raising too many player objections. The objections that were raised at the Aladdin came from the more knowledgable players, the loss of whose business was not necessarily regarded as tragic.

- *Hitting soft 17.* We have previously looked at the reasons why a casino would do this, which has become standard practice across the Las Vegas market. Zender noted that many casino patrons actually favored this change that gives the casino an additional .2% edge, since it gives them a second chance to beat the dealer when they stand on a low hand.

  Example 6.4.7. If the dealer's upcard is a 6, basic strategy calls for a player to stand on 12 or higher. If the player stands on a 13, and the dealer then turns over an ace for a soft 17, a losing hand for the player is turned into a hand with a second chance to win.

- *Dealing $5\frac{1}{2}$ decks from a six-deck shoe, and $1\frac{1}{2}$ decks in a double-deck game.* This was a change that was difficult to sell to floor personnel, d to their overestimate of the threat posed by card counters. Nonetheless, Zender sold it as a way to spend more time dealing and less time shuffling, and the added casino revenue per six-deck table per year due to this switch is in the hundreds of thousands of dollars.

- *Offering late surrender.* While this option gives players a .06% edge over the casino, the fact that few players exercise it, and fewer still use it correctly (Example 6.2.2), makes this a relatively inexpensive rule to implement, and it can be explained as compensation to players for hitting soft 17s. What the casino receives from this rule change is insight that might help identify card counters, since most players who surrender correctly are doing so in conjunction with knowledge of the count.

- *Ignoring card counters whose top wager was $50 or less.* Such counters, whatever their skill level, pose very little threat to the casino's bottom line. This change established the Aladdin as a counter-friendly casino, and numerous aspiring counters—of all skill levels—descended on their games, raising the casino's income. Since the number of truly skilled card counters is wildly overestimated, and the number of novice counters prone to errors is large, this change benefited both parties.

Pessimistic observers confidently predicted that card counters would exploit the new procedures to their fullest advantage and to the detriment of the casino [43]. The reality was far different: The net effect of these changes was that the Aladdin's blackjack revenue during this period was consistently 2% over the average for the Las Vegas Strip [102].

Card Counting In Baccarat

Baccarat, of course, is also a game in which cards are not replaced after each hand, and so prior hands affect future hands. This naturally leads to the question of whether card counting as employed in blackjack can be used in baccarat to improve a player's position relative to the casino. The answer, perhaps surprisingly, is "no." There are a number of reasons for this:

- Baccarat play is rigidly prescribed, with no place for player decisions. Card counting often calls for different player action under different deck compositions, including such actions as doubling down or splitting pairs that are not available in baccarat.

- An important part of the strategy behind card counting in blackjack involves standing on a low hand when the deck is rich in high cards in the hopes that the dealer will bust. Since baccarat hands do not bust if they go over 9, there is nothing to be gained here by tracking specific cards. Ten-count cards in blackjack count zero in baccarat; thus there's no real advantage in tracking how many remain in the shoe.

- There is no bonus for a natural in baccarat, whereas a natural 21 in blackjack pays better than even money (3 to 2 or 6 to 5, depending on the casino). Knowing when natural 8s or 9s are more likely to occur might have some value as far as increasing a wager, but even a natural pays only 1 to 1.

- The bettor may wager on either the Banker or the Player hand. Since the rules are approximately symmetric for the play of the two hands, there are no cards that can be said to be more favorable to one side than the other, as is the case with ten-count cards in blackjack, an excess of which favors the player over the dealer.

Not long after the introduction of card counting methods to blackjack, mathematicians naturally examined how information gained from counting cards might be used to gain an edge at baccarat. Edward O. Thorp soon concluded that "no practical card counting systems are possible" for baccarat [23]. Given a particular block of six cards ready to be dealt, it is possible for the sharp player to gain an advantage. For example, if it is known that the next six cards to be dealt are all 7s, which leads to both hands totaling 4, drawing a third card, and tying at 1, a large Tie bet is guaranteed to win—but it is extremely unlikely that any baccarat player would encounter such a setting. Is it possible to count cards at baccarat? Of course—many card counting systems have been developed. Does it matter? No—the advantage that you might gain is, at best, microscopic. One source claimed a 15¢ advantage per shoe for a player betting $1000 when the shoe turned favorable, an edge of .015%, which is hardly worth the effort or the risk of exposure [33].

Nonetheless, most casinos are happy to provide scorecards to players who wish to track the winning hand from deal to deal in an effort to find and exploit patterns. The patterns certainly exist, but this is only because in an extended gaming session, there will always be apparent patterns in the sequence of winning hands. For all practical purposes, the identity of the winning hand in a round of baccarat is independent of the previous rounds' winners.

Example 6.4.8. In Example 5.3.11, we considered the Banker 9 side bet at baccarat. The dependence of any one hand on the cards previously dealt means that the probability of winning this bet is not fixed. A look at the expectation for the Banker 9 wager shows that if the probability of winning the bet is p, then the bettor has an edge if $10p - 1 > 0$, or if $p > .10$. This suggests that careful tracking of the cards remaining in the shoe might reveal a point where the bet favors the gambler. In light of the fact that there are many ways to draw a two-card total of 9, this might seem like too complicated an undertaking, though, so a simple approach focuses on the most common combination totaling 9: a dealt 9 with a ten-count card.

Thorp and William Walden then proceeded to outline a strategy that would lead to a player advantage for the Banker 9 wager. The strategy involves counting the total number of cards n and the total number of 9s t. In a fresh eight-deck shoe, $n = 416$ and $n/t = 13/1$; as n/t decreases, indicating an excess of 9s over the other cards in the shoe, the Banker 9 bet becomes more advantageous for the gambler. The ratio n/t, combined with the value of determines the optimal amount to wager, as a fraction of a gambler's total bankroll V, on the Banker 9 bet. For example, if $n \approx 130$ and $n/t = 5$, the optimal strategy calls for a wager of 4.7% of V [89].

This was terrific in theory, but was it practical for casino use? A team of players tried the system in two Nevada casinos, winning $100 per hour for seven nights in one and $1000 per hour for two hours in the other before being barred from both. The net result of this research was that Nevada casinos discontinued the Banker 9 side bet [88, p. 180].

6.5   EXERCISES

Answers begin on page 255.

6.1. Consider the following scenario: On the first round of a single-deck black-jack game, there are three players. You are dealt $K\diamond Q\clubsuit$, player #2 is dealt $2\clubsuit 7\spadesuit$, and player #3 is dealt $6\heartsuit 9\spadesuit$. The dealer's upcard is the $3\clubsuit$. You naturally stand on your 20. Player #2 doubles down on her 9 and draws the for a 19. Player #3 stands on his 15, as basic strategy directs. The dealer turns over the $4\clubsuit$ and draws the $K\spadesuit$ for a 17, and so the first hand ends with you and player #2 winning and Player #3 losing.

On the next hand, you have bet $10 and the dealer's upcard is the A You have $A\spadesuit 10\diamond$, a natural 21. Based on this information, is a $5 insurance bet a good idea? Explain your answer by computing the expected value of the total bet including insurance, bearing in mind that if the insurance bet wins, your original bet pushes, and if the insurance bet loses, your original bet pays off at 3 to 2.

6.2. Some casinos have offered an insurance bet when the dealer's upcard is a ten-count card, which pays off at 10 to 1 odds if the hole card turns out to be an ace.

a. Assume that the game is being dealt from a fresh double deck and that the only cards you can see are your hand and the upcard. Calculate the expected value of this bet if you hold

    i. No aces.

    ii. One ace.

    iii. Two aces.

b. If you have a natural, calculate the expectation of your total wager, assuming that you make an insurance bet for half of your original bet and that blackjack pays off at 3 to 2. How does it compare to the expectation without an insurance bet?

6.3. Use the infinite deck approximation to estimate the probability of winning the Lucky Lucky side bet (page 194).

6.4. The Four Queens Casino in downtown Las Vegas once offered a "Red/Black" blackjack side bet on the color of the dealer's upcard [98]. The bet paid off at even money, with the provision that if the upcard was a deuce of the color bet, the wager pushed. Find the expectation of this bet.

6.5. The following hypothetical blackjack situation was first described by Edward Thorp:

> You are playing one-on-one against the dealer, and have determined that the last five cards remaining to be dealt are three 8s and two 7s.

What should your strategy be here? (Hint: Basic strategy may not hold the answer to this endgame opportunity.)

In practice, you would never see a game dealt down this far, nor would you know with 100% accuracy what the last five cards were. Even single-deck games reshuffle before getting down this far into the deck, and shoe games typically reshuffle with 1 to $1\frac{1}{2}$ decks left. Bob Stupak once invented a blackjack variant called Experto 21, which used a single deck and was dealt down to the last card, but one card was burned at the start of the hand precisely to prevent endgame gambits such as this one. To counteract the obvious advantage this offered to a player with a good memory, naturals at Experto 21 paid only even money—which increases the HA by about $2\frac{1}{2}$% [77, p. 31–2].

6.6. Single-deck blackjack under reasonable rules, including a 3 to 2 payment on naturals, is nearly an even game, and for that reason, casinos tend to scrutinize play at single-deck tables more closely. A technique called *depth charging* has been used by some high rollers to try and gain an advantage single-deck games. The gambler plays all seven hands on the table, betting increasing amounts on each hand. For example, the wager on the first hand may be $100, then $200, $300, and so on, up to $700 on the last hand to receive cards. In practice, single-deck blackjack is dealt with all player cards face down, and players playing multiple hands may not look at any hand until all previous hands are completely played out.

How does a player gain any advantage from depth charging?

6.7. Blackjack Switch offers a side bet called the *Super Match* bet, which pays off if the player's first four cards contain a pair or better. The payoff table for the Super Match bet is

Combination	Payoff
One pair	1 to 1
Three of a kind	5 to 1
Two pairs	8 to 1
Four of a kind	40 to 1

Assuming a hand that's being dealt from the top of a fresh six-deck shoe, show that the house advantage on this bet is approximately 2.55%.

6.8. The *Perfect Pairs* side bet, seen at the MGM Grand Casino in Detroit, pays off the following amounts if the player is dealt a pair in his or her first two cards:

- **Mixed Pair**: If the pair contains cards of different colors, the payoff is 5 to 1.

- **Colored Pair**: If the pair consists of two cards of the same color but different suits, the bet pays 10 to 1.

- **Perfect Pair**: A pair consisting of two identical cards (for example, two 7$\diamondsuit$s) pays 30 to 1.

For a blackjack game dealt from a five-deck shoe, calculate the house advantage for Perfect Pairs.

6.9. The Tropicana Casino in Las Vegas once offered a blackjack bonus for a hand consisting of four red 5's [98]. For a game dealt from a six-deck shoe, calculate the probability of being dealt such a hand. How does this probability change in the infinite deck approximation?

6.10. In a four-deck game without surrender, 100 cards have been dealt out, in accordance with the following table:

Low cards:	45
7s:	4
8s or 9s:	13
High cards:	38

On the next hand, you have been dealt 9$\heartsuit$7$\spadesuit$. The dealer shows the 10

a. Compute the high-low count and determine the correct play for this hand.

b. Compute the K-O count and determine the correct play for this hand.

6.11. *Pinkjack* is a variation of blackjack that was offered at the Star Casino in Sydney, Australia [53]. The game's name comes from a bonus payoff for a blackjack with an ace and jack of diamonds, which is called a *pinkjack* pays off at 5 to 1.

a. In a six-deck game, compute the probability of being dealt a pinkjack.

b. An *Ultimate Pinkjack* occurs when both the player and dealer have a pinkjack. In addition to the 5 to 1 pinkjack payoff, this hand carries a fixed bonus payment of $1000. In a six-deck game, what is the probability of an Ultimate Pinkjack?

c. How does the probability of an Ultimate Pinkjack change in the infinite-deck approximation?

6.12. The Instant 18 side bet (page 186) gives a player making it a hypothetical hand with a value of 18. In [31], we find the following distribution for the final value of the dealer's hand in a single-deck game where the dealer stands on soft 17:

x	17	18	19	20	21	> 21
$P(x)$	.1458	.1381	.1348	.1758	.1219	.2836

Using this distribution, find the expectation of a $1 Instant 18 wager.

6.13. Consider a single-deck blackjack game with a full table of seven players, plus the dealer. Given that the average blackjack hand uses up 2.7 cards, find the probability that all four aces will be dealt on the first round, leaving no possibility for a natural in the second round.

6.14. If you're playing single-deck blackjack and are dealt a pair of 7s against a dealer 10, a variation on basic strategy specific to one-deck games states that you should stand on your 14 rather than hitting, as called for in Table 6.3. In light of the fact that the dealer's hole card is more likely to be a 10 than any other value, explain this strategy change.

Betting Strategies: Why They Don't Work

Here's an inescapable mathematical fact:

> *No betting strategy can overcome a negative expectation.*

This is because expected value is additive; that is, $E(X_1 + X_2) = E(X_1$ $E(X_2)$ (Theorem 4.2.1). Translated into English, this means that if the expectation of a bet is negative, making bigger bets, or multiple bets together, will still produce a negative expectation. A more mathematical way of assessing multiple bets is this: The expectation of a collection of simultaneous bets can never be better than the expectation of the most favorable bet or worse than the expectation of the least favorable bet.

This does not stop people from developing and marketing "get rich quick" schemes for pretty much any casino game. In this chapter, we will look at some of the more common ruses either intentionally or unintentionally offered for sale to unsuspecting gamblers.

7.1 ROULETTE STRATEGIES

There are numerous strategies proposed for roulette, perhaps more than for any other casino game. This is likely because roulette is easy to understand and to play, and thus systems for roulette carry the lure of easy money with little effort. By contrast, blackjack gives an impression of requiring considerable effort to master, and the craps layout is complicated and can intimidate new gamblers.

Example 7.1.1. Here's a simple strategy for roulette which was first published in a Cuban magazine, *Bohemia*, in 1959 and further described by gambling scholar John Scarne [60, p. 373].

> *Bet one chip on black and one on the third column, which,*

owing to a fluke of the layout, is unbalanced and contains only four black numbers among its dozen.

The usual claim here revolves around the idea that this bet covers 26 of the 38 slots on the wheel, and 4 of them are covered by both bets. Considered only at that level, this may sound promising, but we need to consider the expected value of the two-unit bet before passing our final judgment:

- If one of the four black numbers in the third column (6, 15, 24, and 33) comes up, both bets win, and we profit $3.

- If one of the eight red numbers in the third column comes up, we the column bet at 2 to 1 but lose the color bet at 1 to 1, for a net w of $1.

- If one of the 14 black numbers in the first or second column comes up, we break even, losing the column bet but winning $1 on the color bet.

- This leaves only 12 numbers—10 red and 2 green—on which we lose

All possibilities are covered here, and so the expected return is

$$E = (3) \cdot \left(\frac{4}{38}\right) + (1) \cdot \left(\frac{8}{38}\right) + (0) \cdot \left(\frac{14}{38}\right) + (-2) \cdot \left(\frac{12}{38}\right) = -\frac{2}{19},$$

and the standard HA of 5.26% has returned. This collection of bets, each of which has an HA of 5.26%, combines to produce a compound bet with the same house edge. Note here that since we bet $2, a return of $-\$\frac{2}{19}$ represents a loss of 5.26% of our original wager.

Needless to say, this was not the mathematical analysis provided by the original purveyor of this scheme.

Example 7.1.2. The *martingale* or *double-up* strategy is rediscovered from time to time, and is most often employed on even-money roulette propositions such as red or black. It can also be used in craps, blackjack, baccarat, or any other game where the probability of winning a bet is close to .5.

> *Bet $1 on the first spin of the wheel. If you win, pocket your $1 profit. If you lose, bet $2 on the next spin. If you win, your net profit is $1—quit while you're ahead. If you lose again, bet $4 on the next spin—win and you walk away with $1, lose and your next bet is $8, and so on. In short, you double up after every losing spin. Since you will win eventually, when you do, you'll walk away a $1 winner.*

Leaving out the concern that this seems like a lot of effort in order to win $1, let's see what we have here. Every statement in this scheme is true until the very end. It is true that you will win "eventually" (the probability of an infinite string of losses is 0), but that can take a very long string of spins. For example, if you bet on red and red fails to come up ten times in a row, an event with probability $\left(\frac{20}{38}\right)^{10} \approx .001631 \approx \frac{1}{613}$ (small, to be sure, but not impossibly so), then your losses total $1023, and your 11th bet would be for $1024 in an attempt to wipe out all of your previous losses and come out ahead. That's a lot to risk for such a small payoff—and you are still more likely to lose than win that bet. If you lose that one, you're down $2047 and are next supposed to put $2048 on the line to come out $1 ahead.

But there's another problem that arises. Casinos typically have maximum bet limits on their roulette tables. For example, suppose that your chosen roulette table displays the sign in Figure 7.1.

ROULETTE $1 GAME

INSIDE BETS	OUTSIDE BETS
$1 MINIMUM	$5 MINIMUM
$100 MAXIMUM	$2000 MAXIMUM
ANY BET	PAYOUT

PLEASE REDEEM ROULETTE CHECKS HERE

FIGURE 7.1: Roulette table sign showing bet limits

This sign refers to two different bet limits:

- On "inside bets," which are bets made on the numbered part of t layout covering from one to six numbers, such as a single-number bet on 23 or a double street bet covering 1 to 6, the minimum bet is $1 and the maximum bet is $100.

- On "outside bets," which are placed along the edge of the layout on even-money and 2 to 1 bets such as red/black or the three colum the minimum is $5 and the maximum *payout* is $2000. The limit on the payout effectively limits even-money outside bets to $2000 and 2 to 1 outside bets to $1000.

The limits are higher for outside bets because the payoff odds are lower, so the casino is exposed to less risk. For the martingale, you will be making outside bets.

If you begin with a $5 bet (the minimum for outside bets) on red, then after nine straight losses, the table limit of $2000 on payouts will prevent you from making the next bet, for $2560, that will recoup all of your losses and give you a $5 profit. Since you're making even-money bets, the maximum of $2000 on payouts is also the limit on your martingale wager.

Suppose that you begin with a $1 bet at a table with a $1000 maximum on outside bets and play this system exactly 1024 times betting on red, where your probability of winning on each spin is approximately .4737. In 485 games, you'll win your dollar on the first spin. On 255 more, you'll lose $1 on the first spin but then win back $2 on the next for a net win of $1. You'll lose on the first two spins before winning on the third in 134 games—again profiting by $1—and so on until game 1022, when you lose the first nine spins, dropping $511 before making it all back with a winning $512 bet on the tenth spin. Through the first 1022 games, you'll be ahead $1 on each, for a total of $1022. On games 1023 and 1024, though, you lose on the first ten spins, for a total loss of $1023 and no opportunity to make the 11th bet that wipes out all your losses. The best you can do at that point is to bet $1000 and hope to win that bet, which will leave you down only $23. Lose that bet, however, and you'll find yourself down $2023 and still unable to bet more than $1000.

A more accurate description of this martingale would be "When you win, you'll win $1 each time. But when you lose—and if you use this strategy ofte enough, you will eventually lose—you'll lose around $1000." The system would be foolproof if you had an infinite bankroll and the casino had no bet limits—although if you had an infinite amount of money, what would be the point of gambling? The thrill of the game, by itself, would seem to be insufficient motivation.

Example 7.1.3. The *Fibonacci* progression attempts to build a system out of the Fibonacci sequence of integers [3]. The first two numbers in the Fibonacci sequence are $F_0 = 0$ and $F_1 = 1$, and each successive term of the sequence is found by adding the two previous numbers together. Thus, the sequence continues $F_2 = 0 + 1 = 1, F_3 = 1 + 1 = 2$, and so on. The sequence begins

$$0, 1, 1, 2, 3, 5, 8, 13, 21, 34, 55, 89, 144, \ldots$$

and can be extended as long as desired.

A roulette system based on the Fibonacci sequence works like the martingale. You begin by betting $1, but then instead of doubling your bet after each loss, you bet the number of dollars given by the next number in the Fibonacci sequence. After you win, instead of returning to a $1 bet, you bet the previous number in the sequence, and so move your way up and down the list as the spins accumulate. Owing to the nature of the sequence, you bet somewhat less than twice your previous bet after you pass a $2 bet; the increase between successive bets after a loss is approximately 61.8% rather than 100%. Nonetheless, this suffers from the same flaw as the martingale—following a stretch of losses, you will be betting a large sum in order to come out a small amount ahead—you'll just lose somewhat more slowly and hit the casino's maximum bet limit after a few more bets than with the martingale. Using the standard martingale strategy with a $1 minimum and $1000 maximum wager allows you to make ten bets before hitting the maximum; the Fibonacci strategy can endure 16 straight losses before the next bet, for $1597, would exceed $1000.

Example 7.1.4. The *Kryptos* system for roulette is a variation on the martingale that attempts to account for the independence of successive wheel spins. The idea behind Kryptos is that there are three attributes to every nonzero roulette number: high/low, odd/even, and red/black. As described in [38],

> *Here, what we are trying to do is forecast, at random, the next spins of the wheel, counting on the fact that we really can't forecast them correctly.*

Six spins of the wheel are recorded, in an effort to secure a random sample. Suppose that those spins are 9 (low, odd, red), 11 (low, odd, black), 33 (high, odd, black), 21 (high, odd, red), 36 (high, even, red), and 6 (low, even, black). We are interested only in the attributes of the numbers, and not the numbers themselves, and so we arrange those attributes in a table:

L	O	R
L	O	B
H	O	B
H	O	R
H	E	R
L	E	B

The argument in support of Kryptos is that it's highly unlikely that all six of these patterns will be completely wrong, thus making it highly likely that betting the even-money chances as listed will result in one win per column by the end of the sixth spin. We now make three separate even-money bets for $1, on low, odd, and red—corresponding to the first row. If a bet wins, we collect our profit and cease betting that column until all six rows are complete. If a bet loses, we double the bet on the choice indicated in that column of the next row (a martingale system). For example, if we make our three bets and the first number then spun is 2 (low, even, and black), we take down of winnings on low and bet $2 each on odd and black (row 2) for the next spin. If that spin yields the high/odd/red 21, we collect $2 from the odd bet and next bet $4 on red. The next spin is a black 11, so our next bet is $8 on red. That spin is a red 19, so we collect $8. Our accumulated losses come to $1 + 1 + 2 + 4 = $8, while our wins total $1 + 2 + 8 = $11—a $3 profit, or precisely the payoff of three separate martingales.

Clearly, the risks of Kryptos are that one or more of the three bets will lose six times in a row, or that the table limit on even-money bets will be reached. The probability of losing six straight even-money bets at American roulette is

$$p = \left(\frac{20}{38}\right)^6 \approx .0213$$

—so there's about a 2% chance that any one column will lose six times in a row and cost $63. While the author of this system claims that "this will be more than offset by our winnings," it will take a lot of $3 winning sessions to make up such a setback.

Example 7.1.5. The *Cancellation* or *Labouchère* system is often suggested for roulette, but it can be used for any game with approximately even-money wagers [60, p. 368].

> If your goal is to win $N, write down a sequence of numbers that add up to N. For your first bet, add the top and bottom numbers together and wager that amount. If you win, cross the numbers out. If you lose, add the amount you lost to the bottom of the list. For the next bet, add the top and bottom numbers and bet that amount, and continue in this manner.
>
> Every time you win, you cross off two numbers. Every time you lose, you add only one. Therefore, you will eventually cross off all the numbers, at which time your profit is the $N you sought.

An immediate problem here is that a gambler using this system will be making increasingly larger bets as the gaming session goes on. Consider the following sequence of wagers and spins, where you are out to win $36:

Original list: {1, 2, 3, 4, 5, 6, 7, 8}

Assume that you always bet on red. The first bet is for 9 units, and loses.

Second list: {1, 2, 3, 4, 5, 6, 7, 8, 9}

The second bet is for $10 and wins. The net profit is $1.

Third list: {2, 3, 4, 5, 6, 7, 8}

The next bet is for $10 and loses. The net loss is –$9.

Fourth list: {2, 3, 4, 5, 6, 7, 8, 10}

The fourth bet is for $12 and wins, for a net win of $3. The list now reads {3, 4, 5, 6, 7, 8}. Note that the amount you have won is equal to the sum of the numbers you have crossed off the original list—the $9, $10, and $12 bets were just means to this end.

Suppose that the next three bets—for $11, $14, and $17—all lose. The list now reads {3, 4, 5, 6, 7, 8, 11, 14, 17}, you are down $39, and your next bet is to be for $20.

If your luck is average or worse, the net effect of the Cancellation System is that you will be betting large amounts in order to win small amounts, for you'll never be ahead more than the sum of the numbers that you originally wrote down. At some point, then, the very real risk that you will run out of money comes into play, and no system works if you can't afford the wager.

An apparently self-published pamphlet called *Famous Las Vegas Gambling Systems* [15] purports to offer winning systems for a range of casino games. Several of the roulette schemes mentioned in this text are simple betting strategies that ignore the fact that successive spins of a roulette wheel are independent:

- Watch the game until either low (1–18) or high (19–36) numbers have failed to come up five times in a row, then start betting that half of the layout (the "cold" or "due" numbers fallacy).

- Going in the other direction: wait until three straight odd or even numbers have come up, and then bet *with* the streak (the "hot" numbers fallacy).

- Wait until one of the columns on the layout has not come up for seven spins, then bet that column ("cold" numbers again).

About the only advantage to be derived from strategies like these is that you'll lose, on the average, less money by virtue of not betting on every spin while you wait for the necessary streak to be established. The author of this brochure does note that these systems do not guarantee a win, but then assures the reader that he has "won consistently" using them.

Example 7.1.6. A betting strategy from *Famous Las Vegas Gambling Systems* that does not rely on disregarding independence calls for repeated betting on one of the columns; these bets pay off at 2 to 1. For an initial bet of the scheme calls for the following:

- On any winning bet ($p = \frac{12}{38}$), take down your winnings and let the original bet ride.

- If the first bet loses, bet $1 again.

- If the second bet also loses, bet $2 on the third spin.

- Follow this "two losing spins, then double your bet" strategy until you win a bet.

The author then states that "you may do this [double your bet after two straight losses] as many times as you like." This is just the martingale approach with a slight delay built in, as you will double a bet only after two consecutive losses. Of course, the 2 to 1 payoff means that you'll be ahead after a win, but the risk of a long string of losses, during which the amount wagered slowly climbs, is greater than with the martingale on an even-money bet.

7.2   CRAPS STRATEGIES

Example 7.2.1. In [60], John Scarne describes the following scam, called th *Watcher* or *Patience System*. The purveyors of this betting strategy either have no understanding of the concept of conditional probability or understand it and are deceiving their customers. Neither of those is a ringing endorsement of their methods.

> *Make $10 at craps almost every time. Simply hold your money until the shooter has made four straight passes, then bet $10 on Don't Pass on the fifth come-out roll. Since the probability of a shooter making 5 consecutive passes is less than 3%, your chance of winning is a whopping 97%!*

Here's what's right about this system:

- The probability of a shooter making five straight points is $(.492)^5$ .0288, which is indeed less than 3%. This is because successive rounds of craps starting with a come-out roll are independent, and so the simple version of the Multiplication Rule (Theorem 3.2.1) can be used.

Here's what's wrong about it:

- The probability we should be considering when we make that Do Pass bet is not "What is the chance of a craps shooter making fiv consecutive points?" but "What is the chance of a craps shooter making 5 consecutive points, *given that he or she has already made four points?*

Since the individual rounds are independent, this probability is just the probability of making a single point, which is .492. By betting on Don't Pass, your chance of winning is 47.9%, less than half the advertised 97%.

The system goes on, however, by adopting a martingale strategy should you lose that first $10 bet. If the shooter makes that fifth point, simply make a $20 Don't Pass bet on the sixth point, and so on. While a win will take you back to a $10 profit, a long losing streak will once again bump up against casino's bet limit.

Like the roulette systems mentioned in *Famous Las Vegas Gambling Systems*, one small redeeming feature of this system is its requirement that you wait until four straight passes have been made before making a bet. This will keep you out of action for long periods of time—the probability of four straight passes is $(.492)^4 \approx .0586$, meaning that you won't be betting on about 94% of all rolls—and thus decrease the rate at which you lose money.

Example 7.2.2. The *Iron Cross* is a combination of craps bets that illustrates Theorem 4.2.1 perfectly. Many sources (see [32] for an example) are honest enough to admit that the overall expectation is still negative.

To bet the Iron Cross, a gambler makes the following four simultaneous bets:

- A $5 Field bet. Recall from Example 4.2.13 that this is a bet that next roll will be 2, 3, 4, 9, 10, 11, or 12. At many casinos, a field bet pays 2 to 1 if a 2 or 12 is rolled; it pays even money otherwise.

- Three Place bets (see Exercise 4.5)—a bet that the number placed will be rolled before a 7—on the numbers 5, 6, and 8. The 6 and 8 place bets pay off at 7 to 6, and so should be made for $6 to avoid rounding in favor of the casino. The place bet on the 5, which pays off at 7 to 5, should be made for $5.

With $22 at risk, Table 7.1 collects the outcomes for each possible roll. The place bets are only resolved if the number placed or a 7 is rolled, and remain active, or "ride," in casino lingo, when any other number is rolled.

TABLE 7.1: Outcomes for the Iron Cross craps bet

Rolls	Field	Place 5	Place 6	Place 8	Net Win	Prob.
2,12	Wins	Rides	Rides	Rides	$10	2/36
3, 4, 9, 10, 11	Wins	Rides	Rides	Rides	$5	14/36
5	Loses	Wins	Rides	Rides	$2	4/36
6	Loses	Rides	Wins	Rides	$2	5/36
8	Loses	Rides	Rides	Wins	$2	5/36
7	Loses	Loses	Loses	Loses	–$22	6/36

Bets that lose are replaced by the bettor. The surface appeal of this bet is easy to see: one of the four bets wins whenever any number other than a 7 is thrown, and thus the Iron Cross returns money 5/6 of the time.

The expected return on each roll of the dice is then

$$E = (10) \cdot \left(\frac{2}{36}\right) + (5) \cdot \left(\frac{14}{36}\right) + (2) \cdot \left(\frac{14}{36}\right) + (-22) \cdot \left(\frac{6}{36}\right) = -\$\frac{14}{36}.$$

Dividing by the $22 wagered gives a house edge of 1.76%.

A modification of this strategy, called the *Unbeatable Iron Cross*, disregards both the additivity of expectation and the law of independent trials. The bets are the same, but the bettor is directed not to place his or her first bet until a point is established, thus avoiding the possibility of losing all of the bets on a come-out roll of 7. Bets are replaced as they lose until the point is made, at which point all bets that remain are left to ride until they are resolved and are not replaced after a loss [32].

The inherent assumption behind the Unbeatable Iron Cross is that a 7 is somehow "due" after the point has been made, and so risking further money is unwise. Anyone who understands that successive rolls of the dice are independent will appreciate the flaw in this reasoning: the probability of rolling the dreaded 7 remains at $\frac{1}{6}$, regardless of any or all previous rolls.

7.3 SLOT MACHINE STRATEGIES

There are literally dozens of suggestions for slot machine players about how to choose a machine or how to choose their wager in an effort to improve their chances of winning or the amount of their winnings. Few of these stand up to mathematical scrutiny.

The fact of the matter is this: With the computer technology that runs a slot machine generating thousands of random numbers every second, no amount of "strategizing" once play has begun will have any more than a random effect on the outcome. This includes such practices as playing rapidly or slowly, opting to pull a handle rather than push a button to spin the reels, and playing without a player's club card out of suspicion that the computer is set to give lower payoffs to club members.

It is not widely appreciated that identical-looking machines may have very different payoff odds, due to the internal settings on the computers running the games. While information about the exact payoff percentage of a given machine is difficult to obtain—for obvious reasons—there are some general principles that are valid. One is that the *hold percentage*—the proportion of the wagered money that is retained by the casino—is generally higher for lower denomination machines, and decreases as the denomination rises. Penny slots may return 88–90% of the money inserted, holding 10–12%, while dollar slot machines might return 95% and hold only 5%, and $5 slots return 95–97.5% [66, p. 45–47]. It must be stressed that these are *long-term* percentages; this does not mean that an individual player will get back 97¢ of every dollar played. Over the course of the machine's life, during which it takes in thousands of dollars, a casino operator can confidently count on a certain percentage being held as profit, but the return to a single player betting a couple of dollars is far more variable.

Example 7.3.1. On a slot machine with a 94% payback percentage, an initial investment of $100 should, *on the average*, return $94. If these proceeds are played back into the machine, the expected return after the second cycle is $100 \cdot .94^2 = \$88.36$. If the player continues "reinvesting" all of his or her winnings, the expected holdings after n cycles is $\$100 \cdot .94^n$—an amount that approaches 0 as n increases.

Short-term fluctuations—in either direction—are likely, but the casino will get its percentage in the long run.

This difference in payback percentages becomes more important when modern video slot machines are considered. The denomination may well be 1 but if a machine offers 20 paylines and the chance to risk 9 credits per line, a player making the maximum bet (which is often easily facilitated by pressing a MAX BET button) is putting $1.80 on the line with every spin of the virtual reels. This simple act effectively turns a penny machine into a $1 machine, but without the more favorable payback percentage of the higher denomination. The intangible factor is, as always, the entertainment value of the 1¢ machine. Modern video slot machines often resemble intricate video games with complex bonus games that are an entertainment experience in themselves. The risk is that a player will invest far more money than he or she intended in a gaming session chasing "just one more bonus round." While good advice to the slot player would be to play on the highest denomination machine your average bet per spin matches, there are benefits to lower-denomination machines that confound stark mathematical analysis.

A strategy that has some validity concerns *progressive* slot machines. These may be identified by the presence of one or more large prize amounts displayed on a video screen above a single machine or a bank of several machines. A typical progressive slot machine operates by extracting a small amount of each bet placed and adding it to the progressive jackpot, which increases with player activity and is awarded under a particularly rare set of circumstances, such as hitting three jackpot symbols on a designated payline with maximum coins bet.

Example 7.3.2. The Lion's Share progressive slot machine at the MGM Grand in Las Vegas is a single machine, the only one remaining of a bank of several originally installed, that has been in place for nearly 20 years. The progressive jackpot is over $2.2 million and though the machine is antiquated by slot technology standards, Nevada law states that it cannot be removed until the jackpot is won. Las Vegas legend holds that the lucky gambler who hits the jackpot, which requires a $3 bet and lining up three lion symbols on the lone payline, will receive the machine as a bonus in addition to the small fortune [46].

Naturally, in light of the jackpot amounts—often running into millions of dollars—these prizes are very rarely won. Less obvious is the fact that, in order to finance the big jackpot, progressive slot machines typically pay off at a lower rate than other slot machines—even nonprogressive versions of the same model. One valid conclusion that can be drawn is this:

> *If you're going to play a progressive slot machine, make sure that you bet enough to activate the conditions for winning the big jackpot.*

Whether this is activating all payoff lines or making the maximum bet per line, you need to do it. It is entirely possible to hit the right combination of symbols but lose out on the big prize because you haven't wagered enough money on that payline. If betting the necessary amount to make the progressive jackpot available exceeds your bankroll or your comfort level, then you need to find a different slot machine.

7.4 BLACKJACK STRATEGIES

Dahl's Progression is a betting system that claimed to supersede card counting and pave the way to player riches [10]. The system calls for a rising sequence of wagers; one version is

$$1, 1, 1.5, 1.5, 2, 2, 3, 3, 5, 5, 7, 7, 10, 10.$$

Using this sequence, a player begins at the first bet in the sequence, wagering that many betting units, so a player at a $5 minimum table would multiply each of the numbers above by 5. If you win a hand, you move to the next number in the sequence, with two exceptions. If you win with a natural, skip one step, and if you win on a double down, skip two steps unless the amount that you would then wager is more than you just won. If you lose, you start over again at the beginning. Should you reach the end of the sequence, stay there and flat bet until you lose.

The appeal of this system is easy to identify: When you're on a losing streak, you're making small bets—always your minimum bet—but as you win, you increase your bets. A long winning streak will leave you with a tidy profit.

As a general rule, though, this system is doomed to fail. The probability of winning 14 straight hands and reaching the end of the sequence above without naturals or double downs is approximately

$$\left(\frac{.42}{.91}\right)^{14} \approx 1.990 \times 10^{-5} \approx \frac{1}{50,245}.$$

This probability considers only resolved hands—ties are not included and might increase the probability slightly—but the message is clear: despite the tales of success related in the book, it is highly unlikely that you will reach the end of this sequence in casual play. Indeed, in practice, you spend a lot of time making minimum bets.

Moreover, Dahl's Progression does not account for double losses arising from split pairs and unsuccessful double downs. If you *lose* two bets at once, which is by no means improbable, there's no way to go back to before the beginning of the sequence and make a bet for less than 1 unit.

Example 7.4.1. Suppose you're at stage 7 in the sequence above, having won six straight hands. You are currently ahead by 9 units and are risking 3 of them on the next hand. If you are dealt a pair of 8s, proper play is to split them regardless of the dealer's upcard. This puts 6 units at risk on a play that is largely defensive: by splitting 8s, you're trying to turn one very weak hand into two middling hands and salvage some of your increased wager, but you have a very real risk of losing 2/3 of your accumulated winnings if you make the correct play, split your 8s, and lose both hands.

That having been said, Dahl's Progression goes against the mathematics and recommends only splitting 8s against a dealer 2 through 8 [10, p. 17–18]). If you choose to defy basic strategy like this and play out your 16, you have a higher probability of losing, but you will only lose 3 units. Which path is better? Many players using this system will opt not to make the mathematically correct play. By promoting a strategy which discourages splitting pairs and doubling down, the progression system is giving away much of the player's flexibility to make larger bets under favorable circumstances or to wager more to improve a bad situation as when splitting 8's, and as a result is increasing the house edge. Clever betting cannot overcome a negative expectation, especially when the betting system increases the magnitude of that negative expectation.

The method behind the apparent madness that is Dahl's Progression may be found in the subtitle of *Progression Blackjack*, the book where it was first published: "Exposing the Card Counting Myth and Getting an Edge in 21." [10] One goal of this book was to discredit card counting as a means to success in blackjack, and once that was done, this system was right there as the way for players to fill the void. A recurring mantra of the progression system is "Card counting is hard. Progression betting is easy." Both statements are basically correct—card counting may not be hard in its own right, but it's certainly harder than progression betting. As is so often the case with simple slogans, though, the truth—in this case, the mathematical truth—is far more complicated.

While it may not be possible for the casual observer to judge the relative merits of progression betting vs. card counting, one fact that cannot be discounted is that casinos are known to bar players whom they suspect are winning by counting cards. Progression bettors are typically welcome at any time, since they are always making wagers where their expectation is negative. This suggests rather strongly that casino management does not consider progression betting as a serous threat.

7.5 AND ONE THAT DOES: LOTTERY STRATEGIES

When we claim that the lottery strategies described in this section work, it's necessary to be clear about what that means. Following the ideas presented here to choose your lottery numbers will *not* change your probability of winning. Any set of, say, six numbers from 1 to 47 has the same probability of winning, whether it is a combination like $\{1, 2, 3, 4, 5, 6\}$ or something more random like $\{7, 20, 25, 35, 43, 45\}$. A life-changing lottery jackpot in Powerball or a similar lottery, when it's won, will be split among all tickets bearing the winning numbers, so your goal in picking numbers is to choose numbers that are unlikely to be chosen by other players. You will still have the same (tiny) probability of winning the jackpot, but if you do win, you will be far less likely to have to share your winnings with another player or players.

The basic idea behind the strategy outlined here is very simple: Find out what other people are likely to do, and *don't do that.* For purposes of illustration, we shall consider a Classic Lotto 47 game, where players pick six numbers in the range from 1 to 47. Studies have shown that many lottery players tend to pick their lotto numbers using some combination of the following methods [29]:

1. Choose numbers with personal significance, such as birthdays, anniversaries, or other important dates.

2. Choose numbers that make interesting patterns on the bet slip.

3. Choose numbers toward the center of the bet slip.

4. Choose numbers in arithmetic progression, such as the set $\{7, 14, 21, 28, 35, 42\}$. Seven, of course, is considered by many to be a lucky number, and so this progression starts with 7 and counts by 7s.

There is some overlap among these methods, of course—numbers in arithmetic progression often make interesting patterns on the bet slip, for example. Avoiding the numbers that these methods tend to choose is the key to minimizing the likelihood that you will have to share the jackpot in the unlikely event that you win it.

The first criterion is the easiest to avoid. Lottery numbers based on birthdays will not be greater than 31, so a player seeking to pick a unique set of numbers should primarily choose high numbers, with no more than one under 31. In practice, we would seek to make the sum of the chosen numbers large.

Consider the general r/s lottery (page 38), where players select r numbers in the range from 1 to s. Classic Lotto 47, then, is a 6/47 game. In an lottery, the sum of the numbers in any possible combination will have (see [29]) mean

$$\mu = \frac{r \cdot (s+1)}{2}$$

and standard deviation

$$\sigma = \sqrt{\frac{(s-r) \cdot r \cdot (s+1)}{12}}.$$

Your goal in choosing numbers should be to have a sum higher than about 75% of all possible sums. Since the sums of the r numbers chosen in an lottery are bell-shaped and symmetrically distributed about the mean, the Empirical Rule (Theorem 4.3.4) can be used here.

Example 7.5.1. In a 6/47 lottery such as Classic Lotto 47, we find that the sum of a player's numbers has mean $\mu = 144$ and SD $\sigma \approx 31.37$. The Empirical Rule states that about 68% of the sums will fall between 113 and 175, and since the data set is bell-shaped, 50% of the values lie below the mean of 144. If we choose numbers that add up to 166 or greater, we will have a combination whose sum exceeds 75% of all possible sums, and we will avoid most popular date combinations.

Of course, it is necessary to take the other factors listed above and explained below into account, for while the combination {42, 43, 44, 45, 46, 47 may have a high sum, it is actually a relatively popular choice and fails our other tests of suitability [29].

Since lottery bet slips are optically scanned, interesting patterns on the bet slip that are not associated with arithmetic progressions tend to be clusters of adjacent numbers. Once again, we look at common practice and then do something else. This factor depends on the exact layout of the betting slip, which in turn may depend on s. A general rule is to avoid too many clusters, which may correspond to arithmetic progressions, and also to avoid too few, which result from bettors making nice-looking patterns. Toward that end, we define the *cluster number* of a wager as the number of adjacent—either edge-adjacent or diagonally adjacent—blocks of numbers. A single number in isolation comprises a cluster, as does any string of bet squares connected at their edges or corners. For an r/s lottery, cluster numbers of 1 or s are to be avoided, as they tend to correlate with smaller overall payouts at the high levels [29, p. 92].

Edge numbers are numbers that appear on an edge of the bet slip, and their exact number and values depend on the format of the slip, as with clusters. Players trying to be "random" in their selections often choose few numbers near the edges, so if you don't want to share your prize, look to the edge numbers in selecting your combination [29]. From a practical perspective, you should select combinations with at least four edge numbers—keeping in mind, of course, the need for a suitably large sum and an appropriate cluster number.

Finally, we address arithmetic progressions. A combination such as {1, 2, 3, 4, 5, 6} is very simple, very popular, and no more or less likely to win than any other [29]. In no small part, this is because the rule for determining this combination is very simple and easily discovered by many people. The same objection may be raised against a combination like {5, 10, 15, 20, 25, 30 or even one like {8, 15, 22, 29, 36, 43}. To avoid such popular methods of choosing numbers, we define the *arithmetic complexity* of a combination:

Definition 7.5.1. The *arithmetic complexity* (AC) of a set of r numbers is the number of positive differences among all of the numbers in that set, minus $r - 1$.

Example 7.5.2. The set {5, 23, 25, 28, 44, 46} chosen by an online random number generator leads to the following positive differences:

$46 - 5 = 41$	$46 - 23 = 23$	$46 - 25 = 21$	$46 - 28 = 18$	$46 - 44 = 2$
$44 - 5 = 39$	$44 - 23 = 21$	$44 - 25 = 19$	$44 - 28 = 16$	
$28 - 5 = 23$	$28 - 23 = 5$	$28 - 25 = 3$		
$25 - 5 = 20$	$25 - 23 = 2$			
$23 - 5 = 18$				

Note that every number is subtracted from every other number in such a way that the difference is positive. There are 12 different differences—2, 21, and 23 are repeated—so the AC of this set is $12 - 5 = 7$.

Example 7.5.3. For the set {1, 2, 3, 4, 5, 6}, the numbers give rise to the following positive differences:

$6 - 1 = 5$	$6 - 2 = 4$	$6 - 3 = 3$	$6 - 4 = 2$	$6 - 5 = 1$
$5 - 1 = 4$	$5 - 2 = 3$	$5 - 3 = 2$	$5 - 4 = 1$	
$4 - 1 = 3$	$4 - 2 = 2$	$4 - 3 = 1$		
$3 - 1 = 2$	$3 - 2 = 1$			
$2 - 1 = 1$				

We count five different differences, so the AC of this set is $5 - 5 = 0$.

When a set of r different numbers is in an arithmetic progression like those above, the AC will always be 0—this is why we subtract $r - 1$. An AC of 0 suggests that the set in question is not very complex, and that certainly applies to an arithmetic sequence. The maximum AC is $r(r-1)/2 - (r -$ when all of the differences are different.

Since lottery players tend toward patterns and simple combinations, the gambler seeking an edge should avoid those and choose more complex sets: anything with an AC of at least 6. This excludes few combinations (in a 6/49 lottery, over 96% of combinations have $AC > 6$ [29]), but they are popular combinations.

Taking everything together, we have arrived at one possible strategy for choosing numbers that will decrease the likelihood of choosing someone else's numbers and thus having to split the prize if you should win: For a 6/47 lottery, play only combinations fitting all four of the following criteria:

- The sum of the numbers is at least 166.

- The cluster number is between 2 and 5.

- The edge number is at least 4.

- The AC is at least 6.

In addition, any combination that satisfies these requirements should nonetheless be discarded if it is a recent winning combination in this lottery or a neighboring state's lottery, for people are known to favor those combinations as well, or if it has a recognizable pattern—for example, {32, 33, 35, 38, 42, 47}.

Another option when buying a lottery ticket is to let a computer select your numbers randomly, an option called "Quick Pick" or something similar. This makes it easier to buy a ticket and may actually be an option worth using if you're trying not to share a prize—provided that your random selection isn't a collection of possible birthday numbers, for example. If you have the chance to examine your Quick Pick numbers before committing to the ticket, this can be an easy way to risk money on a longshot without much likelihood that your numbers will be chosen, nonrandomly, by another gambler.

7.6 HOW TO DOUBLE YOUR MONEY

Q: What's the best way to double your money in a casino?
A: Fold it in half and put it back in your wallet!

All kidding aside, consider the following scenario, a variation of one posed in [27]: You have $500 and need $1000, and no less than $1000, urgently. With an eye toward doubling your money, you enter a casino, determined to bet until you either hit your goal or lose everything—reaching $999 is as useless to you as losing all $500. What strategy gives you the best chance of achieving your goal?

One option is the "big bet" approach: wagering your entire bankroll on one even-money bet and hoping for the best. This has the advantage of being quick—one way or the other, your dilemma will be resolved in about a minute.

It might be argued, however, that this method is riskier than it needs to be in staking everything on one bet, and that you might be better off making a bet with a higher risk but higher payoff, so that you will have several chances to win.

Consider the big bet approach at roulette, where you'll stake all $500 on a single even-money bet. Your chance of doubling your money is $18/38 = .4737$.

If, instead, you put your $500 down at a blackjack table, your chance of winning is about .49 under moderately favorable rules and using basic strategy (see Section 6.3). With that in mind, though, it should be noted that it is unwise to risk your entire bankroll at the start of any one blackjack hand, as this will leave you with no money to split pairs or double down, as you should do when your first two cards give you an edge over the dealer and the rules allow you to get more money on the table.

If you make your one big bet on the pass line at craps, you have a .492 chance of winning and reaching your goal. If you bet the don't pass line, your chance is about .493, allowing for a come-out roll of 12 that does not resolve your initial bet. If you are playing at a craps table that allows free odds bets, it may be possible to devise a betting strategy that allows you to risk less money up front, but this puts you in a position where a win on the come-out roll won't get you all the way to the goal of $1000.

So far, so good. Let's look at spreading the risk (or spreading the opportunity) across more bets. Roulette has a number of options at the same HA, so we'll begin there. If you bet $30 on a single number and it hits once, you can cash out for $1050 and walk away regardless of how many bets you've lost earlier. Five hundred dollars allows for 16 such bets—and all you need to do is win one. The chance of that happening is

$$P(\text{Win at least once}) = 1 - P(\text{Lose 16 out of 16}) = 1 - \left(\frac{37}{38}\right)^{16} \approx .3473$$

about 12.5% *less* than making a single even-money bet.

By switching to a street bet on three numbers, your money won't last for quite as many spins, but your chance of winning on any spin is tripled. A bet will give you five chances at a $990 payoff, and since your bet is returned with your win, your total after a single win will exceed $1000. The chance of winning one of those 5 bets is

$$P(\text{Win at least once}) - 1 - P(\text{Lose 5 out of 5}) = 1 - \left(\frac{35}{38}\right)^{5} \approx .3371,$$

which is not as good as any previous option.

A trend seems to be developing here. Let's look at the other choices for roulette bets. In each case, the amount of the bet has been rounded to a convenient number.

Bet	Payoff	# of bets	Wager	P(Win at least 1)
Straight	35 to 1	16	$30	.3473
Split	17 to 1	8	$60	.3511
Street	11 to 1	5	$90	.3371
Corner	8 to 1	4	$125	.3591
Basket	6 to 1	3	$166	.3451
Double street	5 to 1	3	$166	.4028
Dozen	2 to 1	1	$500	.3684
Even-money	1 to 1	1	$500	.4737

One note: If you're going to make a dozen bet, your first bet can be for only $250, since a win there gives you $500 in winnings and the return of your wager. Taken together with the $250 you didn't bet, you have $1000. However, if your first bet loses, you then are faced with raising $1000 starting with only $250—a somewhat more daunting task than the original.

The conclusion is clear: If you're trying to double your money at roulette, your best chance to do so is by making a single even-money bet.

The reason for this is very simple and has nothing to do with roulette: The more bets you make, the greater the opportunity for the house advantage, whatever the game or the edge may be, to work against you. The phrase "Go big or go home" is highly appropriate to this quest, and this applies regardless of the game you're playing. If your tastes run more toward slot machines, you're better off making your bets on the highest denomination machine you can find, because the rate of return is typically higher on higher-denomination slot machines, and fewer spins will be necessary to win the money you seek. Find your way to the high-limit room of your favorite casino. If there's no $ machine available (they exist, and pay out as much as 1000 to 1 on certain reel combinations) then you should seek out the highest denomination they do provide.

Example 7.6.1. A bet with multiple payoffs, such as the Field bet at craps, might suggest an alternate strategy, since the possibility of a better-than-even payoff provides the option of making a smaller initial bet and holding back some money for a second wager. If you bet $250 on the first roll, there's a 1/18 chance that a 2 or 12 will be rolled, bringing your total immediately to $1000; if you lose that bet, you still have $250 for another chance at reaching $1000.

Suppose that you bet $250 on each roll until you have at least $1000 or run out of money. Let $\hat{p} = P(E)$ be the probability of reaching or exceeding your $1000 goal. The following disjoint events are part of E:

- E_1: A 2 or 12 is rolled on the first roll, and you have $1000. The probability of this event is

$$p_1 = \frac{2}{36}.$$

- E_2: You win even money on two consecutive rolls, bringing your total to $1000. This event has probability

$$p_2 = \left(\frac{14}{36}\right)^2 = \frac{196}{1296}.$$

- E_3: You win even money on the first roll and 2 to 1 on the second roll for a total of $1250.

$$P(E_3) = p_3 = \frac{14}{36} \cdot \frac{2}{36} = \frac{28}{1296}.$$

- E_4: You lose the first roll, win 2 to 1 on the second, and win either 1 to 1 or 2 to 1 on the third roll. Your total is either $1000 or $1250, and the probability of this event is

$$p_4 = \frac{20}{36} \cdot \frac{2}{36} \cdot \frac{16}{36} = \frac{640}{46,656}.$$

- E_5: In the first two rolls, you win one and lose one, returning you to a balance of $500. When your balance is $500, your probability of eventually winning is $\hat{p}$. Since there are two different orders (win-lose and lose-win), we have

$$p_5 = 2 \cdot \frac{14}{36} \cdot \frac{20}{36} \cdot \hat{p} = \frac{560}{1296} \cdot \hat{p}.$$

- E_6: You lose on the first roll, win 2 to 1 on the second to bring your total to $750, and lose the third roll to return you to $500, from which your chance of winning is again $\hat{p}$.

$$P(E_6) = p_6 = \frac{20}{36} \cdot \frac{2}{36} \cdot \frac{20}{36} \cdot \hat{p} = \frac{800}{46,656} \cdot \hat{p}.$$

Adding everything up gives

$$\hat{p} = p_1 + p_2 + p_3 + p_4 + p_5 + p_6 = \frac{353}{1458} + \frac{655}{1458} \cdot \hat{p},$$

which is a linear equation in $\hat{p}$ whose solution is

$$\hat{p} = \frac{353}{803} \approx .4396.$$

This is still inferior to the probability of .492 of doubling your money by making a single pass line wager.

7.7 EXERCISES

Answers begin on page 256.

7.1. More recently, the roulette system published in *Bohemia* (Example 7.1.1) has been called the "Three-Two System," and calls for a bet of 3 units on black and 2 units on the third column [97]. Find the expectation of this version of the bet.

7.2. The *d'Alembert* system for roulette is said to be the work of French mathematician Jean le Rond d'Alembert. Its fundamental premise is an excellent illustration of the Gambler's Fallacy. The description here (from [3]) assumes that you're betting on red:

> *Since red and black occur equally often, you simply increase your bet by one unit after each loss and decrease it by one unit after each win until you get back to your original starting bet, and then bet one unit each spin until you lose.*

Apart from its embrace of incorrect mathematics in its disregard for independence, find the flaw in this system.

7.3. Here's a roulette system posted on the Internet in July 2012:

> *Beginning with a bankroll of $450, wager on a single number until either it wins or you go broke after 50 straight losses. On spins 1–20, wager $5 per spin. On spins 21–40, wager $10 per spin. On spins 41–50, wager $15 per spin.*
> *As soon as your number hits, you quit and leave with a profit.*

Assume that you're playing American roulette.

a. Find the probability that your number will hit exactly once in 50 spins.

b. By computing your total holdings if your number comes up for the first time on spins 20, 40, and 50, show that you will indeed be ahead if your number hits once.

c. Find the probability that you will lose all of your initial bankroll.

7.4. Absent the potential for a 2 to 1 payoff when a 2 or 12 is rolled, doubling your money on a Field bet would be a simple "one-shot" problem. Show that the probability that you will double your money from $500 to $1000 on a single Field bet is greater than the probability of .4396 calculated in Example 7.6.1.

7.5. The *Red Snake* strategy at roulette has been imbued by some with mystical powers. Like the *Bohemia* magazine strategy described in Example 7.1.1, it relies on a pattern in the roulette layout. To make the Red Snake bet, a

Appendix A: House Advantages

Wager	House advantage	Page #
Blackjack, single-deck, Las Vegas Strip rules	~ 0.00%	199
Craps, pass line bet with 1000X odds	.0014%	148
Craps, pass line bet with 4X odds	0.28%	147
Super Bowl XLVII coin toss proposition bet	0.98%	120
Baccarat, Banker bet with 5% commission	1.06%	94
Baccarat, Player bet	1.23%	94
Craps, don't pass or don't come line	1.34%	84
Craps, pass or come line	1.41%	84
European roulette, all bets	2.70%	82
Pai Gow Poker	2.84%	114
Three card poker, Q64 as beacon hand	3.37%	113
Three card poker, call on queen or higher	3.45%	113
Let It Ride	3.50%	109
American Royal Roulette, all bets	4.00%	140
American roulette, Colors bet	4.34%	139
Sports betting, one game	4.55%	119
Caribbean stud poker	5.22%	105
American roulette, all bets except basket bet	5.26%	82
Crapless craps, pass line bet	5.40%	145
Craps, field bet	5.56%	85
Chuck-a-luck	7.87%	83
American roulette, basket bet	7.89%	83
Craps, hardway bet on 6 or 8	9.09%	86
Blackjack, Royal Match side bet	10.90%	192
Big Six wheel, bet on $1	11.11%	91
Craps, hardway bet on 4 or 10	11.11%	87
Baccarat, Tie bet	14.05%	94
Sic bo, bet on 4 or 17 (correct payoff)	15.28%	154
Diamond Roulette, eight numbers color bet	15.79%	141
Double Action Roulette, single number, both wheels	16.83%	144
Craps, Any Seven bet	16.67%	84
Craps, Fire Bet	24.90%	149
Keno, Meskwaki Casino Super 20 Special	34.60%	89
Keno, Mark 7 bet	36.30%	88
Michigan State Lottery Daily 3 straight bet	50.00%	90
Michigan State Lottery Daily 3 box bet	50.20%	90
Rupert's Island Draw, 4 of a kind side bet	78.20%	160
You Pick 'Em Treasury ticket	99.13%	91

Appendix B: Mathematical Induction

How do you climb a ladder? Reduced to its most basic steps, climbing a ladder requires two simple actions:

1. Climb from the ground to rung #1.

2. Once you are on rung #n, climb up to rung #$n + 1$. Repeat this step as needed.

The proof technique known as *mathematical induction* is essentially this ladder-climbing algorithm. Induction is a useful proof technique for proving certain results about the natural numbers $\mathbb{N}$, and since we shall have occasion to use it from time to time in setting the mathematical foundation for our work in probability, a brief discussion is in order here.

Let $S(n)$ be a statement about the integer n; for example, $S(n) =$ "The sum of the first n integers is equal to $\dfrac{n \cdot (n+1)}{2}$." Formally, mathematical induction proceeds along the following three-step path. Suppose that we seek to prove that the statement $S(n)$ is true for all $n \in \mathbb{N}$. We do the following:

1. **Verify** that the statement $S(1)$ is true—that is, confirm that we can get on the first step of the ladder. This is often called the *base case*.

2. **Assume** that the statement $S(k)$ is true for some integer k, or for all integers $\leq k$. Since we have already shown the truth of $S(1)$, we know that there is at least one value of k where this assumption is valid. This step is known as the *induction hypothesis*. We are assuming here that we are on the kth step of that ladder.

3. **Prove** that $S(k + 1)$ is true, using the assumption that $S(k)$ is true in that proof. This is the "You're on rung k, can you get to rung $k + 1$?" step of the proof.

If these three steps are completed, we have done the following

- In step 1, we confirmed that $S(1)$ is true.

- In light of steps 1 and 3, we can be certain that $S(2)$ is true.

- This result, together with another reference to step 3, tells us that S is true.

- And so on—each step in the process allows us to access the next step, just like climbing a ladder. We conclude that, for any natural number n, $S(n)$ is true, which was the goal of the proof.

We shall illustrate mathematical induction with several examples in which all of the steps are identified.

Proposition B.1. For any $n \in \mathbb{N}$:

$$\sum_{i=1}^{n} i = 1 + 2 + \cdots + n = \frac{n \cdot (n+1)}{2}.$$

Proof.   1. **Verify**: If $n = 1$, then the left side of this expression is just 1. The right side is $1 \cdot (1+1)/2 = 2/2 = 1$, so our first step is complete.

2. **Assume** that the statement is true for $n = k$:

$$1 + 2 + \cdots + k = \frac{k \cdot (k+1)}{2}.$$

3. To **prove** the statement for $n = k+1$, we seek to show that

$$1 + 2 + \cdots + (k+1) = \frac{(k+1) \cdot (k+2)}{2}.$$

Beginning with the left side, we have the following:

$$1 + 2 + \cdots + (k+1) = 1 + 2 + \cdots + k + (k+1)$$
$$= (1 + 2 + \cdots + k) + (k+1)$$
$$= \frac{k \cdot (k+1)}{2} + (k+1),$$

by the induction hypothesis. Continuing, we have

$$1 + 2 + \cdots + (k+1) = \frac{k \cdot (k+1)}{2} + \frac{2(k+1)}{2}$$
$$= \frac{(k+1) \cdot (k+2)}{2},$$

which is the desired result, completing the proof.

Proposition B.2. The sum of the first n odd integers is n^2; that is,

$$1 + 3 + 5 + \cdots + (2n - 1) = n^2.$$

Proof. 1. **Verify:** If $n = 1$, both sides of the equation evaluate to 1, and our first step is complete.

2. **Assume** that the statement is true for $n = k$:

$$1 + 3 + 5 + \cdots + (2k - 1) = k^2.$$

3. **Prove** the proposition for $n = k + 1$: Show that

$$1 + 3 + 5 + \cdots + [2(k + 1) - 1] = (k + 1)^2.$$

On the left-hand side (LHS) of this target equation, we have

$$
\begin{aligned}
\text{LHS} &= 1 + 3 + 5 + \cdots + [2(k + 1) - 1] \\
&= 1 + 3 + 5 + \cdots + 2k - 1 + [2(k + 1) - 1] \\
&= (1 + 3 + 5 + \cdots + 2k - 1) + [2(k + 1) - 1] \\
&= k^2 + [2(k + 1) - 1], \text{ using the induction hypothesis.} \\
&= k^2 + 2k + 1 \\
&= (k + 1)^2 \text{ by factoring.} \\
&= \text{RHS, completing the proof.}
\end{aligned}
$$

Induction can also be used to prove inequalities, as in the following example.

Proposition B.3. For all $n \geq 4$, $2^n < n!$.

Proof. 1. **Verify:** Since the proposition starts at $n = 4$, our base step also begins there. If $n = 4$, then $2^4 = 16 < 24 = 4!$, and the first step is verified.

2. **Assume** that the statement is true for $n = k$: $2^k < k!$.

3. To **prove** the statement for $n = k+1$, we must show that $2^{k+1} < (k+!)!$. Remember that $n > 4$, so $k + 1 > 2$. We have

$$
\begin{aligned}
2^{k+1} &= 2 \cdot 2^k \\
&< 2 \cdot k!, \text{ by the induction hypothesis.} \\
&< (k + 1) \cdot k! \\
&= (k + 1)!,
\end{aligned}
$$

completing the proof.

Finally, we use induction to give an alternate proof of Theorem 2.4.2 (page 27).

Theorem 2.4.2. If $\#(\mathbf{S}) = n$, then there are 2^n events that may be chosen from $\mathbf{S}$.

Proof.   1. **Verify.** If $\#(\mathbf{S}) = 1$, then we may write $\mathbf{S} = \{x\}$. $\mathbf{S}$ then has $2 = 2^1$ subsets: $\emptyset$ and $\{x\} = \mathbf{S}$ itself.

2. **Assume** that a set $\mathbf{S}$ with k elements has 2^k subsets.

3. To **prove** the statement for $\#(\mathbf{S}) = k + 1$, we need to show that $\mathbf{S}$ 2^{k+1} subsets. Write $\mathbf{S} = \{x_1, x_2, \ldots, x_k, x_{k+1}\}$. Consider the set $\mathbf{T}$ $\{x_1, x_2, \ldots, x_k\}$ obtained by removing x_{k+1} from $\mathbf{S}$. Since $\#(\mathbf{T}) =$ it follows from the induction hypothesis that $\mathbf{T}$ has 2^k subsets. Each of these is also a subset of $\mathbf{S}$, since $\mathbf{T} \subset \mathbf{S}$.

   By adding the removed element x_{k+1} to each of the 2^k subsets already found, we generate 2^k new subsets of $\mathbf{S}$ that are not subsets of $\mathbf{T}$. Adding these to the previous subsets gives a total of $2^k + 2^k = 2 \cdot 2^k = 2$ subsets of $\mathbf{S}$, completing the proof.

Appendix C: Internet Resources

Gambling mathematics is one field where it's difficult to separate theory and practice without diminishing the experience. Reading about these games of chance and the odds behind them is one thing, but studying the games is no substitute for playing them and seeing the equations spring to life. The following Web site is a fine resource for free practice versions of casino games, both the traditional and the more obscure. As a bonus, some games will let you know if you're not following proper strategy, and the right strategy is described elsewhere on the Web site. The games here are free to play; while it may be argued that the risk of losing or winning money adds an additional level of realism to the study of casino mathematics, beginners would do well to practice the games with no money at stake. This is also an excellent way to confirm our findings about betting systems without the risk of the big losses that those systems possess.

- **The Wizard of Odds: Free Casino Games**
  http://wizardofodds.com/play/

YouTube, of course, abounds with gambling strategy videos, and Michael Shackleford, the Wizard of Odds behind the gaming Web site above, has produced some which explain the rules and strategies for many casino games. They can be accessed through links at this site:

- **The Wizard of Odds: Instructional Videos**
  http://wizardofodds.com/video/

Answers to Odd-Numbered Exercises

Chapter 1

Exercises begin on page 9.

1.1. $R \cap O = \{1, 3, 5, 7, 9, 19, 21, 23, 25, 27\}$.

1.3. $R \cap H \cap E = \{30, 32, 34, 36\}$.

1.5. $L \cap H = \emptyset$.

1.7. $(H \cup R)' = \{0, 00, 2, 4, 6, 8, 10, 11, 13, 15, 17\}$.

Chapter 2

Exercises begin on page 48.

2.1. 3.075×10^{-4}.

2.3. 8/51.

2.5. .0506.

2.7. .0211.

2.9. 6.4×10^{-11}.

2.11. 190.

2.13. .1961.

2.15. Let r be the number of matched red numbers, w the number of matched white numbers, and $P(r, w)$ the probability of matching r red and white numbers.

r	w	$P(r, w)$
2	2	1/106,525
2	1	48/106,525
1	2	48/106,525
2	0	276/106,525
0	2	276/106,525
1	1	2304/106,525
1	0	13248/106,525
0	1	13248/106,525

Chapter 3

Exercises begin on page 72.

3.1. .3679.

3.3. .0455.

3.5. .0353.

3.7. P(Win an Over 7 bet) $\approx$ .4167. By symmetry, this is also the probability of winning the Under 7 bet.

3.9a. $\frac{9}{38}$.

3.9b. $\frac{9}{38}$.

3.11. P(Win) = .15625, P(Lose) = .0625.

3.13. With 16 barred, P(Win) $\approx$ .5184 > .5. With 2 and 16 barred, P(Win) $\approx$ .5027 > .5.

Chapter 4

Exercises begin on page 134.

4.1a. $P(7) = \frac{1}{6}$.

b. $4/36 = 1/9$.

c. Careful counting of the possibilities will reveal that the probability distribution of the sum of a pair of Sicherman dice is the same as that for a pair of standard d6s.

4.3. $E \approx -.1111$. The HA is 11.11%.

4.5. $E = -\frac{1}{11}$. Dividing by the $6 bet gives an HA of 1.515%.

4.7a. For the straight bet, the expectation is –$.50, and the HA is therefore 50%. For a wheel bet, we have E = –$12. Once again, the HA is 50%.

4.7b. Fo an n-way boxed bet with net winnings of x, we have

n	x	E	HA
24	207	–$.5008	50.08%
12	415	–$.5008	50.08%
6	832	–$.5002	50.02%
4	1249	–$.5000	50.00%

4.9. No five-card poker hand can contain both a full house and a four-card royal flush.

4.11. For a bet on the $2 spot, $E \approx -\$.1667$.

For a bet on the $5 spot, $E \approx -\$.2222$.

For a bet on $10, $E \approx -\$.1852$.

For a bet on $20, $E \approx -\$.2222$.

For a bet on either one of the two logos, $E \approx -\$.2407$.

4.13. –$.01625.

4.15. .2461.

4.17. –$.125, and so the HA is 12.5%.

4.19. 10.94%.

4.21. –$.1078, so the house advantage is 10.78%.

Chapter 5

Exercises begin on page 176.
5.1. 14.3% for either bet.
5.3. 36.1%.
5.5.

Rolls	p	X	E
2, 12	1/7	45	$-4/49 \approx -.0816$
3, 11	2/8	14	$-1/8 = -.1250$
4, 10	3/9	7	$-1/9 \approx -.1111$
5, 9	4/10	5	$-1/25 = -.0400$
6, 8	5/11	3.5	$-17/242 \approx -.0702$

5.7. .1227, so the HA is therefore approximately 12.3%.
5.9a. 0. (Yes, this bet is fair!)
5.9b. −$.08.
5.9c. Bet only on Blue.
5.11. −.0540, −.0789.
5.13. $E \approx -.0278$.
5.15. For the first game, the HA is 29.24%. For the second game, it's 28.96%. The second bet has a lower house advantage—but not by much.

Chapter 6

Exercises begin on page 217.
6.1. The true count is −4. You should not make the insurance bet.
6.3. 2.845×10^{-5}.
6.5. The correct strategy, as outlined by Thorp in [88], is to bet as much as you can on the next hand. Borrow money if you have to, for you are a guaranteed winner.

You and the dealer will each be dealt a total of 14, 15, or 16, with one card remaining to be dealt. You stand, and the dealer must draw. If the dealer has 14, the remaining card must be an 8, and the dealer busts. If the dealer's total is 15 or 16, the remaining card—whether 7 or 8—is a certain bust card. Enjoy your win.

6.9. $\dfrac{1}{782,382}$.

In the infinite deck approximation, $p \approx 2.188 \times 10^{-6}$.
6.11a. 7.42×10^{-4}.
6.11b. 3.87×10^{-7}.
6.11c. 5.47×10^{-7}.
6.13. .0270.

Chapter 7

Exercises begin on page 240.

7.1. $-\frac{5}{19}$.

7.3a. .3562.

7.3c. .2636.

7.7. −$1.

7.11a. .086.

7.11b. 1.17×10^{-5}.

7.11c. .4042.

Bibliography

[1] 2by2—Prizes and Odds. Online at http://www.powerball.com/2by2/ 2by2_prizes.asp.

[2] Andersen, Ian, *Burning the Tables in Las Vegas*, 2nd edition. Huntington Press, Las Vegas, 2003.

[3] Anonymous, *Casino Confidential*. Quirk Books, Philadelphia, 2008.

[4] Baldwin, Roger R., Wilbert E. Cantey, Herbert Maisel, and James P. McDermott, The Optimum Strategy In Blackjack. *Journal of the American Statistical Association* **51**, September 1956, p. 429–439.

[5] Bernstein, Peter L., *Against The Gods: The Remarkable Story of Risk* John Wiley & Sons, New York, 1998.

[6] The Bone Man, *New Craps Fire Bet Emerging in Las Vegas*. Online at http://www.nextshooter.com/firebet.

[7] Brokopp, John, *Gaming Company Makes Roulette More Colorful*. Online at http://brokopp.casinocitytimes.com/article/gaming-company-makes-roulette-more-colorful-58988.

[8] Catlin, Donald E., *A Really Hard Hardway Bet*, in *Finding the Edge* Olaf Vancura, Judy A. Cornelius, and William R. Eadington, editors. Institute for the Study of Gambling and Commercial Gaming, Reno, NV, 2000, p. 297–302.

[9] Craps layout.png, online at http://en.wikipedia.org/wiki/File:Crapslayout.png.

[10] Dahl, Donald, *Progression Blackjack: Exposing the Card Counting Myth and Getting an Edge in "21."* Citadel Press, New York, 1993.

[11] Emert, John, and Dale Umbach, *Inconsistencies of "Wild-Card" Poker* CHANCE, 9:3, 17–22, 1996.

[12] Epstein, Richard A., *The Theory of Gambling and Statistical Logic*, revised edition. Academic Press, San Diego, 1995.

[13] Epstein, Richard A., *The Theory of Gambling and Statistical Logic*, second edition. Academic Press, San Diego, 2013.

[14] *EZ Baccarat: Game Math.* Online at http://www.deq.com/en/table-games/ez-baccarat-math.php.

[15] *Famous Las Vegas Gambling Systems*, Castindes Books Tijuana, n.d.

[16] Ferguson, Tom W., *The Curse of the Big Lottery Win.* Online at http://www.lastchaser.com/BugleTwo.html#anchor_141, 30 September 2012.

[17] Film8ker, American roulette table layout.gif. Online at http://commons.wikimedia.org/wiki/File%3AAmerican_roulette-table_layout.gif.

[18] Film8ker, American roulette wheel layout.gif. Online at http://commons.wikimedia.org/wiki/File:American_roulette-wheel_layout.gif.

[19] The Football Pools, online at http://www.footballpools.com/cust.

[20] Frey, Richard L., *According to Hoyle.* Fawcett Crest Books, Greenwich, CT, 1970.

[21] Gelman, Andrew, and Deborah Nolan, You Can Load a Die, But You Can't Bias a Coin. *The American Statistician* **56**, #4, November 2002, p. 308–311.

[22] Griffin, Peter, *Extra Stuff: Gambling Ramblings.* Huntington Press, Las Vegas, 1991.

[23] Griffin, Peter, *The Theory of Blackjack*, 6th edition. Huntington Press, Las Vegas, 1999.

[24] Galilei, Galileo, *Sopa le Scoperte Dei Dadi.* Online at http://www.leidenuniv.nl/fsw/verduim/stathist/galileo.htm.

[25] Griffin, Peter, and John M. Gwynn, Jr., *An Analysis of Caribbean Stud Poker*, in *Finding the Edge*, Olaf Vancura, Judy A. Cornelius, and William R. Eadington, editors. Institute for the Study of Gambling and Commercial Gaming, Reno, NV, 2000, p. 273–284.

[26] Hadsall, Joe, *No Dice: Casino Invents Version of Craps Played with Cards.* Joplin *Globe*, Joplin, MO, 16 April 2010. Online at http://www.joplinglobe.com/enjoy/x1687713220/No-dice-Casino-invents-version-of-craps-played-with-cards.

[27] Haigh, John, *Taking Chances: Winning with Probability*, 2nd edition. Oxford University Press, Oxford, 2003.

[28] Hannum, Robert C., and Anthony N. Cabot, *Practical Casino Math*, 2nd edition. Institute for the Study of Gambling and Commercial Gaming, Reno, NV, 2005.

[29] Henze, Norbert, and Hans Riedwyl, *How to Win More: Strategies for Intcreasing a Lottery Win.* A K Peters Ltd., Natick, MA, 1998.

[30] How, Stephen, *Card Craps (+EV).* Online at http://discountgambling.net/playcraps/.

[31] Humble, Lance, and Carl Cooper, *The World's Greatest Blackjack Book* Doubleday and Company, Garden City, NY, 1980.

[32] *The Iron Cross Craps System.* Online at http://www.crapscasinos.org/craps-systems/the-iron-cross/.

[33] Jacobson, Eliot, *Card Counting n Baccarat.* Online at http://www.worldgameprotection.com/the-catwalk/casino-ology/BOSVIEW/Card-Counting-in-Baccarat/.

[34] Jacobson, Eliot, *Contemporary Casino Table Game Design.* Blue Point Books, Santa Barbara, CA, 2010.

[35] Jensen, Marten, *Video Poker and Slots for the Winner.* Cardoza Publishing, New York, 2002.

[36] Kolata, Gina, In Shuffling Cards, Seven is Winning Number, *New York Times*, Jan. 9, 1990.

[37] Krigman, Al, *Beware of Rule Variations: Little Things Can Mean a Lot.* Online at http://krigman.casinocitytimes.com/article/beware-of-rule-variations-little-things-can-mean-a-lot-60995, 3 September 2012.

[38] *The Kryptos System for Roulette*, in *The Big Winner's Systems Book* Gambling Times, Inc., Hollywood, CA, 1981.

[39] Las Vegas Advisor, *Question of the Day April 7, 2009.* Online at http://www.lasvegasadvisor.com/qod.cfm?qid=1656.

[40] Lea, Mike, *Modern Progression Systems.* Online at http://www.bjmath.com/bjmath/progress/prog2.htm.

[41] Levinson, Horace C., *Your Chance To Win.* Farrar and Rinehart, Inc., New York, 1939.

[42] *Master Library.* Texas Instruments, Inc., Dallas, TX, 1979.

[43] Meadow, Barry, *Blackjack Autumn.* Huntington Press, Las Vegas, 1999.

[44] Mezrich, Ben, *Bringing Down the House: The Inside Story of Six M.I.T. Students Who Took Vegas for Millions.* The Free Press, New York, 2002.

[45] Mezrich, Ben, *Busting Vegas: The MIT Whiz Kid Who Brought the Casinos to Their Knees.* William Morrow, New York, 2005.

[46] MGM Lion's Share Progressive. Online at
http://vegasslotcritic.com/?p=90.

[47] Michigan State Lottery, *Where The Money Goes*. Online at
http://www.michiganlottery.com/where_the_money_goes?cid=p1core-
tbx09.f.1800/b59a4/a8/
ddbfb883.1d0e9ebe75e0486e4a4a4791fba1d1b5.

[48] Nevada State Gaming Control Board, *Rupert's Island Draw*. Online at
http://gaming.nv.gov/approved_games_rules/Ruperts_Island_Draw.pdf.

[49] New Jersey Casino Control Commission, *Six Numbers Color Wager in
Roulette*. Online at
http://www.state.nj.us/casinos/actreg/reg/docs_proposed_regs/p080707-
2.diamondroulette.pdf.

[50] New Jersey Casino Control Commission, *TEMPORARY RULES: New
Color Wagers in Roulette*. Online at
http://www.state.nj.us/casinos/actreg/reg/docs_temp_regs/t080521.
diamondroulette.pdf.

[51] Nolan, William I., *The Facts of Baccarat*. Gambler's Book Club, Las
Vegas, 1976.

[52] Pai Gow Strategy. Online at http://www.casinoguide.com/pai-gow-
poker-strategy.html.

[53] Pinkjack. Online at
http://www.star.com.au/sydney-casino/Documents/guides/
STARGAMEGUIDE_Pinkjack.pdf.

[54] Player, B.J., *Here's How to Even up the Odds at Roulette*, in *The Big
Winner's Systems Book*. Gambling Times, Inc., Hollywood, CA, 1981.

[55] Roberts, Stanley *et al. The Gambling Times Guide to Blackjack*. Gambling Times Inc., Fort Lee, NJ, 2000.

[56] Roulette Table Layouts. Online at
http://www.ildado.com/roulette_table_layout.html.

[57] Royal Roulette Pty Ltd, *Royal Roulette*. Online at
http://www.royalroulette.com.

[58] Rubin, Max, *Comp City: A Guide to Free Casino Vacations*, 2nd edition.
Huntington Press, Las Vegas, 2001.

[59] Scarne, John, *Scarne on Dice*, 8th revised edition. Crown Publishers,
Inc., New York, 1980.

[60] Scarne, John, *Scarne's Complete Guide to Gambling*. Simon and Schuster, New York, 1961.

[61] Schlesinger, Don, *Blackjack Attack*, 2nd edition. RGE Publishing, Oakland, CA, 2000.

[62] Schoenberg, Frederic Paik, *Introduction to Probability with Texas Hold'em Examples*. CRC Press, Boca Raton, FL, 2012.

[63] Schwartz, David G., *Big Six: A Longitudinal Micro Study*. Las Vegas, Center for Gaming Research, University Libraries, University of Nevada-Las Vegas, 2011.
Online at http://gaming.unlv.edu/reports/big6_study.pdf.

[64] Schwartz, David G., *Roll the Bones: The History of Gambling*. Gotham, New York, 2006.

[65] Schwartz, David G., *Why Double Action Roulette Will Catch On*
Online at http://vegasseven.com/blogs/felt/2012/07/05/why-double-action-roulette-will-catch.

[66] Scoblete, Frank, *Break the One-Armed Bandits*. Bonus Books, Chicago, 1994.

[67] Shackleford, Michael, *Big Six—Wizard of Odds*. Online at http://wizardofodds.com/games/big-six/.

[68] Shackleford, Michael, *Blackjack Rule Variations*.
Online at http://wizardofodds.com/games/blackjack/rule-variations/.

[69] Shackleford, Michael, *Blackjack Switch*. Online at http://wizardofodds.com/games/blackjack/switch/.

[70] Shackleford, Michael, *Craps Side Bets*. Online at http://wizardofodds.com/games/craps/appendix/5/#firebet.

[71] Shackleford, Michael, *Probabilities in Bingo II*. Online at http://wizardofodds.com/games/bingo/probabilities/2/.

[72] Shackleford, Michael, *San Diego Roulette*. Online at http://wizardofodds.com/games/roulette/san-diego-county/.

[73] Shackleford, Michael, *Three Card Poker*. Online at http://wizardofodds.com/games/three-card-poker/.

[74] Shampaign, Charles E., *Handbook on Percentages*. Gambler's Book Club, Las Vegas, 1976.

[75] *The Shotwell System*, in *The Big Winner's Systems Book*. Gambling Times, Inc., Hollywood, CA, 1981.

[76] Sic bo table.png. Online at http://en.wikipedia.org/wiki/File:
Sic_bo_Table.png.

[77] Sklansky, David, and Alan E. Schoonmaker, *DUCY?: Exploits, Advice,
and Ideas of the Renowned Strategist*. Two Plus Two Publishing LLC,
Henderson, NV, 2010.

[78] Smith, Jeff, *Fred Dakota Founded Native American Casinos—In
A U.P. Garage*. Traverse, Northern Michigan's Magazine, October
2009.   Online   at   http://www.mynorth.com/My-North/Octobe
2009/Fred-Dakota-Founded-Native-American-Casinos-In-A-UP-
Garage/index.php?cparticle=1&siarticle=0#artanc.

[79] Smith, John, *No Limit: The Rise and Fall of Bob Stupak and Las Vegas'
Stratisphere Tower*. Huntington Press, Las Vegas, 1997.

[80] Spanish Blackjack. Online at http://www.casinogaming.com/tutorials/
tablegames/spanish21.html.

[81] Stuart, Lyle, *Casino Gambling for the Winner*, revised edition. Lyle
Stuart Publishing, Secaucus, NJ, 1980.

[82] Stuart, Lyle, *Lyle Stuart on Baccarat*. Kensington Publishing Corp.,
New York, 1997.

[83] Stupak, Bob, *Yes, You Can Win!* Galaxy Publishing, Las Vegas, 1992.

[84] Su, Francis E. et al. "Perfect Shuffles." *Math Fun Facts*. Online at
http://www.math.hmc.edu/funfacts.

[85] Tamblin, Henry, *Casino Gambling: The Best of the Best*, 2nd edition.
Research Services Unlimited, Greensboro, NC, 1998.

[86] *Thoroughbred Report: 2010 Triple Crown*. Online at
http://www.thoroughbredreport.com/triple_crown_2010.html.

[87] Thorp, Edward O., *Beat the Dealer: A Winning Strategy for the Game
of Twenty-One*. Blaisdell Publishing Co., New York, 1962.

[88] Thorp, Edward O., *Beat the Dealer: A Winning Strategy for the Game
of Twenty-One*, second edition. Vintage Books, New York, 1966.

[89] Thorp, Edward O., and William Walden, *A Favorable Side Bet in
Nevada Baccarat*. Journal of the American Statistical Association
#314, Part 1 (June 1966), p. 313–28.

[90] *Three Card Poker, 6 Card Bonus: The Fresh(ly Minted Milliona
Maker*. Online at
http://www.vegaschatter.com/story/2012/7/13/183528/145/vegas-
travel/Three+Card+Poker%2C+6+Card+Bonus%3A+The
+Fresh(ly+Minted+Millionaire)+Maker. Published 13 July 2012.

[91] Tilton, Nathaniel, *The Blackjack Life*. Huntington Press, Las Vegas, 2012.

[92] Trucksess, Joseph R., *Table Slots Game and Method of Play*. US patent #6089569, July 2000. Online at http://www.freepatentsonline.com/6089569.html.

[93] Uston, Ken, *Ken Uston on Blackjack*. Barricade Books, Fort Lee, NJ, 1986.

[94] Vancura, Olaf, and Ken Fuchs, *Knock-Out Blackjack*. Huntington Press, Las Vegas, 1998.

[95] *Video Poker FAQ: Questions and Answers*. Online at http://www.pokerheroes.com/video-poker-faq.html.

[96] Walsh, Audley V., *John Scarne Explains Why You Can't Win: A Treatise on Three Card Monte and Its Sucker Effects*. Gambler's Book Club, Las Vegas, 1972.

[97] White, Jeff, *The Three Two Roulette System*. Online at http://www.silveroakcasino.com/blog/online-roulette/three-two-roulette-system.html.

[98] Wong, Stanford, *Basic Blackjack*. Pi Yee Press, Las Vegas, 1995.

[99] Wong, Stanford, *Blackjack Secrets*. Pi Yee Press, Las Vegas, 1993.

[100] Wong, Stanford, *Casino Mistakes—Part II*, in *American Casino Guide* 1999 edition, Steve Bourle, editor. Casino Vacations, Dania, FL, 1999, p. 213–215.

[101] Wong, Stanford, *Wong On Dice*. Pi Yee Press, Las Vegas, 2005.

[102] Zender, Bill, *Casino-ology: The Art of Managing Casino Games*. Huntington Press, Las Vegas, 2008.

[103] Zender, Bill, *Casino-ology 2: New Strategies for Managing Casino Games*. Huntington Press, Las Vegas, 2011.

Index

Printed in the United States
by Baker & Taylor Publisher Services